Landform and Terrain

The Physical Geography of Landscape

Colin Mitchell & Patrick Mitchell

B
Brailsford

First published in Great Britain 2007 by Brailsford with WritersPrintShop

ISBN 1904623565

Cover illustration: Mai Chau, North East Vietnam

Designed by e-BookServices.com

Landform and Terrain

The Physical Geography of Landscape

Colin Mitchell & Patrick Mitchell

Table of Contents

List of Figures

Figure 10.1 Screes, Atlas mountains , Morocco
Figure 10.2 Wastwater screes, Lake District
Figure 10.3 Rock creep, soil creep, and mudflow
Figure 10.4 Slump flow

Figure 11.1 Formation of arêtes
Figure 11.2 Mountain horns, Switzerland
Figure 11.3 A Bergschrund
Figure 11.4 Mountain area after glaciation
Figure 11.5 Mt Snowdon
Figure 11.6 The Seven Sisters Falls, Norway
Figure 11.7 Edinburgh crag and tail
Figure 11.8 Continental glacial deposits

Figure 12.1 Moraines
Figure 12.2 Terminus of a glacier
Figure 12.3 Form of glacial deposits
Figure 12.4 Glacial till
Figure 12.5 More continental glacial deposits
Figure 12.6 Drumlin
Figure 12.7 Drumlin field
Figure 12.8 Kame formation
Figure 12.9 Formation of a delta
Figure 12.10 Stone polygons, garlands stripes

Figure 13.1 Base level and stream grades
Figure 13.2 A meandering river
Figure 13.3 Types of alluvial deposits
Figure 13.4 Creation of an ox-bow
Figure 13.5 Rainbow Bridge, Utah
Figure 13.6 Stream braiding, Death Valley
Figure 13.7 Section of an alluvial fan, Oman
Figure 13.8 Desert landforms
Figure 13.9 The Nile Delta
Figure 13.10 River terraces
Figure 13.11 London: alluvial terraces

Figure 14.1 Cracking clay, Iceland
Figure 14.2 Evaporating salt surface, Jordan
Figure 14.3 Spring mound, Ethiopia

Preface

The word "terrain" refers to the surface characteristics of the ground that surrounds us. Terrain forms the skeleton of our landscape, underlying our homes and gardens and ultimately giving us our food, clothing, and building materials. We see it daily on our way to work. We view it from the plane window as we travel abroad. We seek its charms when on holiday - the trails we walk, the beaches we lie on, and the mountain views that entrance us.

The object of this book is to explain the reasons for the forms and variations of terrain so that we may have a fuller understanding and appreciation of its practical values and attractions. These fascinate the rambler, cyclist, and climber but also the driver and casual visitor.

The title of this book is Landform and Terrain: the Physical Geography of Landscape. These terms have related and somewhat overlapping meanings. We use the term *terrain* for the bare surface of the ground, not including vegetation cover. By *landform* we mean the character of the land surface that results from geology and the surface processes or wind, water and ice. *Landscape* refers to the overall character of a geographical area which may include several different types of terrain and landform. Our aim has been to bring together geology, soil, climate and hydrology in such a way as to give the observer an overall view of the landscape they sees around them.

We have used simple black and white illustrations, often making drawings from photographs. We have acknowledged the sources of all material used. Every effort has been made to contact organizations and individuals for permission to reproduce material and we apologize for any that have been omitted. Where there is no attribution the illustrations derive from photographs or drawings of the authors.

CHAPTER 1

Introduction: The surface we see

Terrain is ground surface, the solid skeleton we stand on. It underlies the covering of vegetation and human artifacts. It is not permanent but always changing. Most changes happen so slowly that we can only see their effects after many years or even centuries. Sometimes change is sudden. Volcanoes and flash floods can make drastic alterations in hours.

How has terrain evolved to its present state? The basis is rock structure. Just as the skeleton determines the contours of the human body, so rocks govern the form of the earth's surface. Just as we cannot understand the body without knowing of the bones, we cannot understand terrain without appreciating rock structures and the processes which modify them. Forces deep in the earth's crust determine its main outlines. Lithology governs the broad surface form. Ice, wind and water mould it. Together they explain why one slope is steep and another gentle, why a valley takes a particular direction, or why granite and limestone give distinct hill shapes. These in turn explain the locations of soils, plants, and human settlements.

No single classification system is ideal for all landforms so we use different systems to bring out the most important features. We make a fundamental distinction between those mainly suffering erosion, which are called "degradational" and those mainly experiencing deposition, which are called "aggra-dational". The former are generally classified by rock type (li-thology), the latter according to their chief agent of deposition: gravity, ice, water, wind, or organic accumulation.

It sometimes happens that a feature we describe under one heading could equally well be described under another. Examples are weathering patterns, valley head forms, and

cavernous weathering features under hard surfaces collectively called *tafoni*. All of these can occur on a number of different
rock types. Where a particular feature is not found under one
rock type in which it most often occurs, it may appear under
another.

The perception of landscape depends on the eye of the be-
holder. Farmers, engineers, and historians, for instance, will
have different reactions to the same scene. The same person
will view it differently when working than when on holiday.

Landscape appeals when it conveys drama, especially when
viewed from a high elevation and where it includes mountains, rivers, lakes, or coasts. It has much emotional attraction.
It can be overwhelming, as when we first see the Matterhorn
or the Grand Canyon and has inspired artists, poets and mu-
sicians from ancient time. It has been an irresistible draw to
many mountaineers and explorers. When we look at a view we
tend to see it somewhat as we do a picture, confined laterally
by vertical limits and moving outwards to the distance from a
foreground to a background enclosing the horizon. Beauty involves contrast between the two. Other such comparisons also
help: between vertical and horizontal, light and dark, straight
lines and curves, land and water, soil and sky. Our eyes tend
to prefer an absence of stiff formal angles and undue repetition
of the same elements. In general the best colour combinations
are those which jump one neighbouring shade in the rainbow
spectrum but not two, so that red goes with yellow, orange
with green etc.

English has a wide geographical vocabulary but needs to
add many borrowed foreign words. This is particularly true of
terms used to describe landform and terrain. They often come
from the local names for significant or unusual phenomena.
Partly for this reason landform study has an especially rich lan-
guage, illustrated by a glance at the index of this book. Some
terms, such as "obsequent fault line scarp" or "recumbent fold"
are both complex and unique.

The abundance of geological photographs which have become available of recent years has revolutionized our appreciation of landscapes but although fast and accurate, they have
some disadvantages compared to drawings. They have a more
restricted field of view and do not discriminate between fea-

tures of different importance to the observer. Before photography was generally available, geologists relied on drawings, which they frequently drew with remarkable skill. Outstanding examples of such skill are William Morris Davis' diagrams of the erosion cycle, Sir Archibald Geikie's illustrations of his textbooks such as *The Scenery of Scotland Viewed in Connection with its Geology*, and Walter Holmes' drawings of the American West, one of which is shown in **Figure 4.1.**

Most examples of terrain given in the book are from Britain. These islands contain within a small compass and fairly uniform climate a remarkable variety of geological structures, rock types, and evolutionary stages. Except for its unusually extensive relics of Ice Age glaciation, Briain would be a virtual microcosm of the humid temperate zone. It is exceptionally well documented in a near-infinity of books, scientific papers, maps, guides, and pictures. But other sources are needed. An especially good aid in studying terrain is the scenery of arid lands where the relations between rock type and landform are most starkly expressed. In this book the aim has been to quote examples which are representative, accessible, and aesthetically pleasing.

Terrain appears differently when viewed from different vantage points. At ground level we are especially conscious of relief, but our view is limited. A higher point of observation gives a panorama. An aeroplane gives a comprehensive view which can be a wonderful help in understanding the ground. In recent years aerial and satellite photography have given synoptic views of wider areas invaluable in recording landscape and relief.

Perceptions of the landscape are also altered by motion. Close-up features pass more quickly than far-off ones so that the faster we travel the more we tend to focus on more distant objects.

Though gradual changes occurring over several miles or more are more easily appreciated from a car or train than on foot, most of the details necessary to a full understanding of a location cannot be assimilated at speed. Despite the wide availability of aerial photography, the essential and enjoyable tools of terrain study remain boots and binoculars.

Chapter 2

Degradational landforms

Degradational landforms consist of exposed rocks whose surface configuration is mainly due to erosion. They tend to form the highest parts of the landscape and coastal headlands. Many varieties are visible in Britain.

Rock type

Rocks reflect their origin and this governs their surface appearance. The primary distinction is between those which have solidified from liquid rock and those that have been laid down as sediments. The first group includes rocks that are still largely as they were when they first solidified from molten lava called *igneous*. Within the sedimentary group the main distinction is based on variations in their hardness in resisting erosion. *Metamorphic* rocks are either igneous or sedimentary that have been substantially altered by the forces of mountain building and vulcanism. Within the igneous and metamorphic groups the main distinction is between the presence or absence of clear directionality in their surface forms. This gives a natural fourfold classification of the resulting landforms: *compact coherents, slaty coherents, compact crystallines* and *slaty crystallines*.

Compact coherents include the calcareous and siliceous sediments, mainly limestones and sandstones. These give landforms reflecting the inclination of their strata and the effects of erosion in differentiating beds of varying resistance.

Slaty coherents include shales and mudstones. They are less resistant to erosion, called lacking *competence*. Their mountains are lower and have less dramatic contrasts. Ridges can have a worn and ragged outline and the landscape appears dreary and desolate.

Compact crystallines include all igneous rocks. These are solidified from molten *magma* (molten rock). Those thrown out on the surface are volcanic, those formed underground are plutonic. On the whole both give bold, smooth, rounded, and non-directional landforms.

Slaty crystallines include the metamorphic rocks. These initially formed by the cooling of molten rock and were then substantially altered by volcanic action, movements in the earth's crust, the chemical deposition of new rock or dissolution of the original. Post-cooling processes can impart intense directional distortions, causing the rocks to show one-dimensional linear structures called *lineation*. The main types of metamorphic rock are gneiss, schist, and phyllite. All tend to give sharp and aligned landforms. In the Alps they weather into pinnacles and splintered ridges. They form many of the most conspicuous peaks, including Mont Blanc, the Matterhorn, and the Point des Ecrins.

Rock structure

There are four main forces that have shaped the Earth's surface and built mountains: vulcanism, faulting, folding, and circumdenudation. Usually more than one of these processes is present at the same place, sometimes all are.

Vulcanism is the ejection of magma from the earth in solid or liquid form. It generates landforms which reflect the size and shape of the vents through which it reaches the surface, the nature of the magma, and the form of the land on to which it is ejected. The forms are considered in chapter 3.

When rocks are folded they have a *dip* and *strike*. The direction of dip is down the steepest part of the bed. The strike is the compass direction along a horizontal line on the plane of the dip **(Figure 2.1A)**.

A *fault* is a crack in rock caused by forces driving it apart **(Figure 2.2A)**. Most faults occur in groups termed *fault zones* or *fracture zones*. If the direction of movement of one side of the fault relative to the other is up or down it is called a *dip fault*, the respective sides being the *upthrown block* and the *downthrown block*. The plane of the fault may be vertical or at an angle. If the latter the fault may be *normal* **(Figure 2.2B)** or

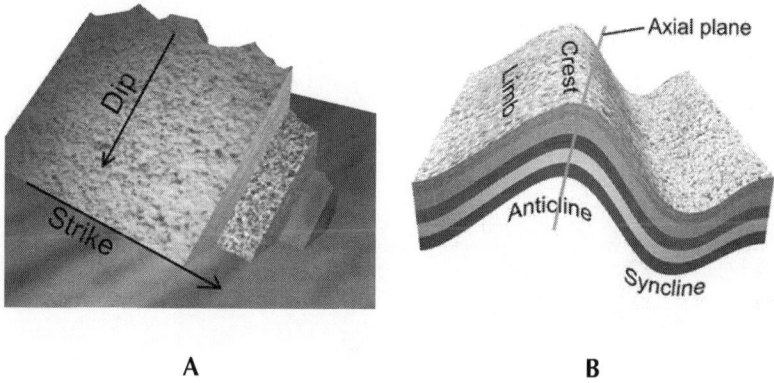

A   B

Figure 2.1: Structural features in sedimentary rocks. With tilted beds the angle they make to the horizontal is known as the dip. The compass direction at right angles to the dip is known as the strike (2.1A). When sediments have been distorted by tectonic action after their deposition synclines and anticlines are formed (2.1B)

reversed **(Figure 2.2C)**. The fault's angle of dip (analogous to the dip of sedimentary strata) is called the *hade*, measured in degrees from the vertical. The amount of vertical displacement is called the *throw*. There is usually a small near-vertical cliff on the upthrown side. This is called the *hanging wall* and its underground continuation the *foot wall*.

We can see visible evidence of some faults at the surface. *Fault scarps* are *outcrops* (surface appearances) of the upthrown block and indicate the position where the faulting originally occurred. Sometimes backwearing erosion moves the scarp back from its original position above the fault. In this event the new scarp is called a *fault-line scarp*. Where the rock on the upthrown side is more easily eroded than that on the downthrown side, erosion can reduce the former to a lower elevation than the latter, reversing the step caused by the fault in the first place. The resulting scarp is known as an *obsequent fault-line scarp*. If the situation is reversed and the downthrown rock is the less competent, the result is a *resequent fault-line scarp*.

Geological folds result from lateral compression and can have many forms, as indicated in **Figures 2.1B and 2.3**. Beds

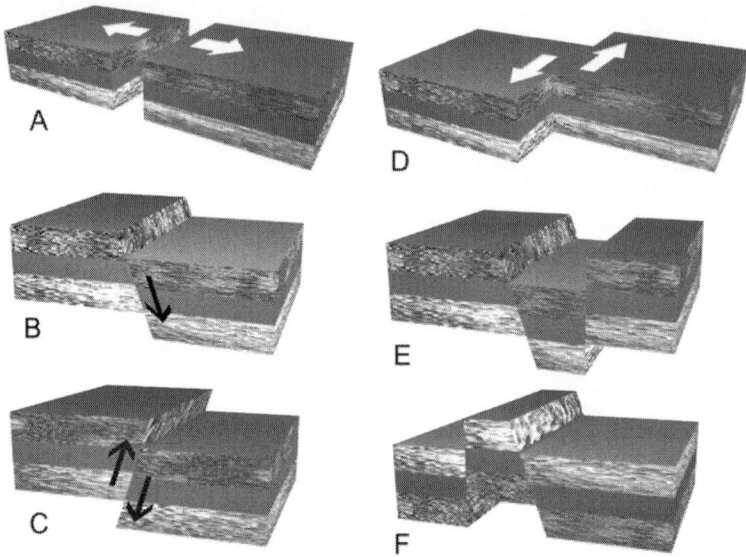

Figure 2.2: Types of fault: A fractured; B normal; C reversed; D tear fault; E graben; F upthrown block.

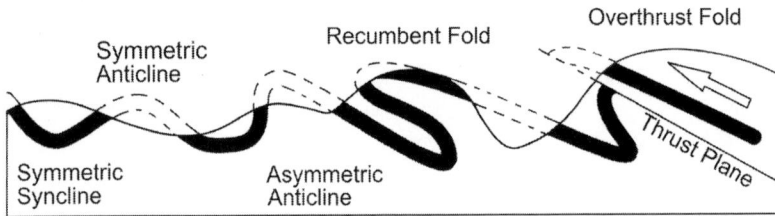

Figure 2.3: Various types of folding and thrusting are illustrated in this hypothetical geological section.

that are simply tilted in predominantly one direction are called monoclines. When a series of tilted beds appear at the surface they have a *scarp slope* where the edges of the beds are exposed and a *dipslope* down the uppermost bed. The resulting landform containing these two features is a *cuesta*. Neighbouring cuestas are separated by *vales*. A landscape consisting mainly of the two is called a *cuesta-and-vale landscape*.

When a series of beds are downfolded into a basin shape they form a *syncline*, when upfolded into a dome, an *anticline* **(Figures**

2.1B and 2.6). When compound these are called respectively *syn-clinoria* and *anticlinoria* **(Figure 2.4)**. They can be *symmetrical* if slopes on both sides are equal or *asymmetrical* if one side is steeper than the other. When bevelled by erosion, these landscapes give rise to characteristic patterns of scarp orientation: anticlines and domes with inward-facing scarps, synclines and basins with outward-facing scarps. A section southeastwards from the Chilterns to the Channel Coast of England illustrates this, the London Basin being an asymmetric synclinorium, the Weald a somewhat symmetric anticlinorium **(Figure 2.5)**. One can see this by comparing the gentle dip of the Chilterns with the steeper dips of both North and South Downs. Greater pressure can form *overfolds* which become *recumbent* folds when more or less horizontal **(Figure 2.3)** and *nappe structures* when at largest scale. Looked at three-dimensionally, circular or oval upfolds and downfolds are known as *domes* and *basins* respectively. When the uplift or depression is centred around a point rather than along a line, the result is a *pericline,* which can be either a dome or a basin. This has an angle of lateral pitch as well as the angle of folding.

Circumdenudation means the erosion of surrounding materials to leave an upstanding core, often large enough to be a mountain. Such isolated mountains are called *monadnocks* whose survival is due to their competence. A small difference in this quality may make a large difference in the topography as can be seen where an area of shale or clay is partly capped or impregnated with siliceous material. Erosion will leave this standing proud of the rest.

Rock competence

Landscapes reflect the variations in erosion-resistance of the different rocks they contain, as well as the local climate and vegetation. Generally the warmer and wetter the climate and the more luxuriant the vegetation the greater the amount of weathering. More competent rocks form bolder landforms or more prominent headlands than their less competent neighbours.

Rock competence is a many-faceted quality that depends on more than simple hardness. Why does chalk form hills and promontories but clay form valleys and bays? Why do the land-

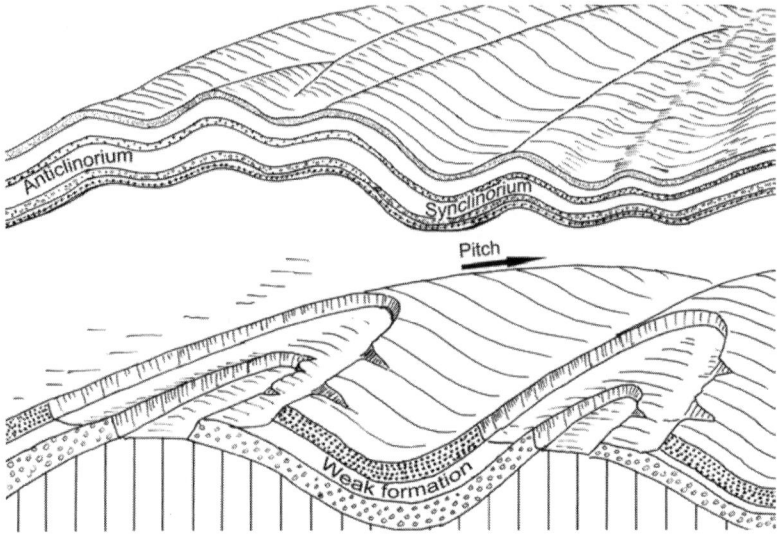

Figure 2.4: Upper part: anticlinorium and synclinorium
Lower part: eroded pitching anticlines and synclines, where the structure is not only folded but the folds are dipping, exposing them to erosion (after Lobeck page 602).

forms on two sandstones differ as much as does the Millstone Grit from the Lower Greensand?

The most vulnerable rocks tend to be dark because they absorb more heat from the sun and are generally richer in chemically alkaline ferromagnesian minerals and lower in the more acid quartz. Certain rocks have particular chemical vulnerability. This is true of calcium carbonate which is subject to dissolution especially in the presence of carbonic acid from plants, and also of aluminosilicate minerals, such as feldspars, which are vulnerable to hydrolysis by the silicates in soil water.

Permeability reduces erosion rates by allowing rain water to soak into the rock rather than run along the surface in a destructive torrent. This is the main reason that soft chalk and sandstone have more surface competence than shale **(Figure 2.7)**. On the other hand permeability also increases the internal dissolution of soluble rocks leading to the formation of caves and hollows.

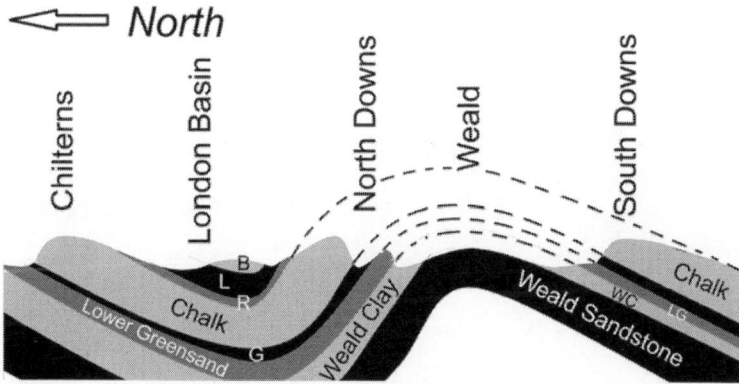

Figure 2.5: Section through southern England to show asymmetrical folding.

Figure 2.6: A simple anticline in limestone near Torre Pellice, Italy.

The single most significant determinant of erosion rate in otherwise similar rocks is probably their cracking properties. Rocks that have few cracks are the most competent. Also, the narrower and the less continuous the cracks the less the elements can penetrate into them and attack the surrounding rock. Cracks that go inward from a rock face rather than parallel to it lead to a reduced tendency for spalls to fall away and may have less effect on competence. On the other hand, this protective effect is reduced because the inward inclination of cracks allows more water to penetrate into the rock face and fewer spalls means less protective mantle of loose rock.

Competence is visible at all scales. A rock which has certain forms over a few metres, or even over a few centimetres, will display similar characteristics where its outcrop covers many kilometres, as illustrated in southeastern England **(Figure 2.8)**.

Drainage texture

Drainage texture is the density of drainage lines per unit area. Where there are many it is fine, where few it is coarse. There are four main governing factors: rock competence, the size of loose surface particles, the overall slope of the land and the thickness of vegetation cover. The finest drainage textures occur where the rocks are incompetent, yield clay-sized particles, where the overall slope of the drainage catchment is steep, and where there is little or no vegetation to bind the soil. Surveyors quantify drainage texture in terms of kilometres of stream length per

Figure 2.7: Sandstone interbedded with shale near Whin Sill, Northumberland.Note the way that the shale is more easily eroded leaving the more prominent sandstone strata (lighter colour) proud.

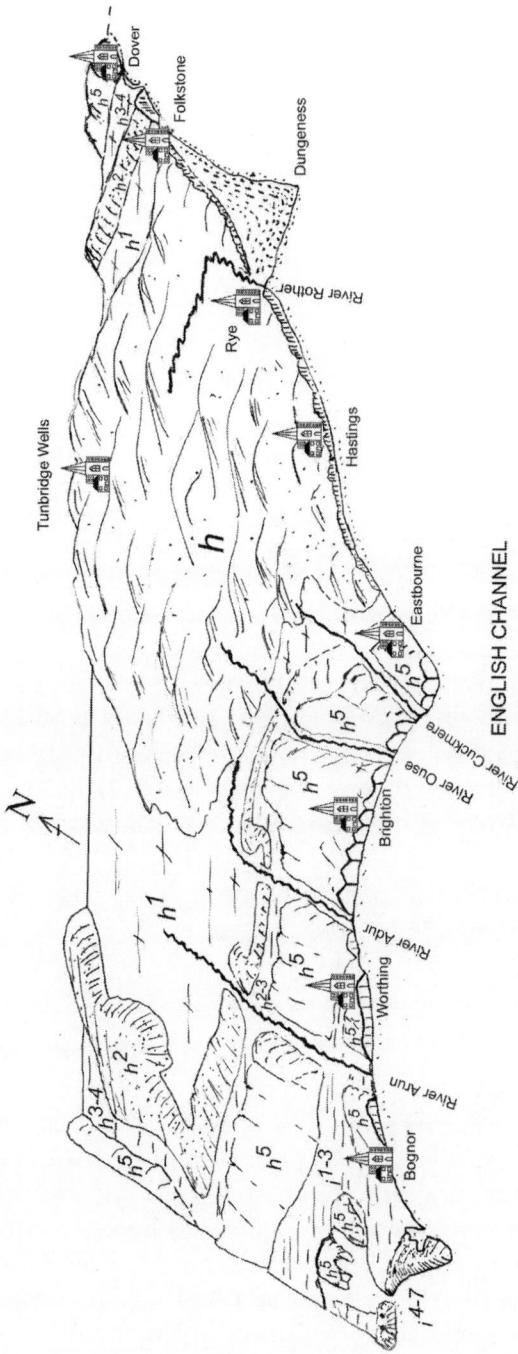

Figure 2.8: Oblique diagram of the Wealden area of southern England. The letter symbols are those used by the British geological survey for the different beds: i4-7 Barton, Bracklesham & Bagshot beds (sandy and gravelly); i1-3 London Clay; h5 Chalk; h3-4 Gault Clay; h2 Lower Greensand; h1 Weald Clay; h Hastings beds (sandstone). Note how erosion has un-roofed the anticline. The result has been to expose the dipslopes of older and older beds northwards from the Channel until one reaches the High Weald around Tunbridge Wells. Northwards from this we begin to re-encounter the escarpments of the same sequence in reverse order.

square kilometre of terrain. Different catchments show wide contrasts in this value: 3-4 for the sandstones of Exmoor and the Appalachian Plateau of eastern USA, 20-30 for the scrub-covered Coast Ranges of California, 200-400 for the Badlands of the Dakotas, and up to 1300 for unvegetated clays. So clear are these relationships, especially when viewed on aerial or satellite photographs, that drainage texture can often be taken as a key indicator of rock type.

Causes of very steep slopes

The presence of very steep slopes in hard rocks is almost always due to one of five causes, or a combination of them:

1. Climate at high altitudes.

At high altitude there is high insolation (sunlight falling on the surface), buffeting by high winds, exposure of the rock to repeated freezing and thawing, and sculpturing by snow and ice. Detritus falls as soon as it is dislodged, so there is no chance for soil to accumulate or for vegetation to gain a foothold. The resulting features are sharp peaks, separated by *divides* (*arêtes*). If the peaks are particularly sharp and steep, they are *aiguilles* or *jorasses*; if they are less so, *horns*. Where the arête is cut by a deep trough caused by overflow from a melting ice sheet, it is called a *col channel*. Below the arête are steep slopes, locally mantled with snow at upper levels and with scree at lower levels **(Figure 2.9)**.

2. Faulting.

Faulting can lead to vertical or horizontal displacements, the latter sometimes being called *tear faults*. A valley between two fault slopes is a *graben* **(Figure 2.2E)**. The opposite situation: an upland delimited by two parallel faults is a *horst* **(Figure 2.2F)**. Both may be simple or compound. Fault lines, when exploited by streams, can make steep gorges. These can be especially deep and narrow when the rock is soluble such as limestone or dolomite (a mixture of calcium and magnesium carbonates), because the gorge is deepened by solution of the underlying rock and the collapse of caverns it may contain.

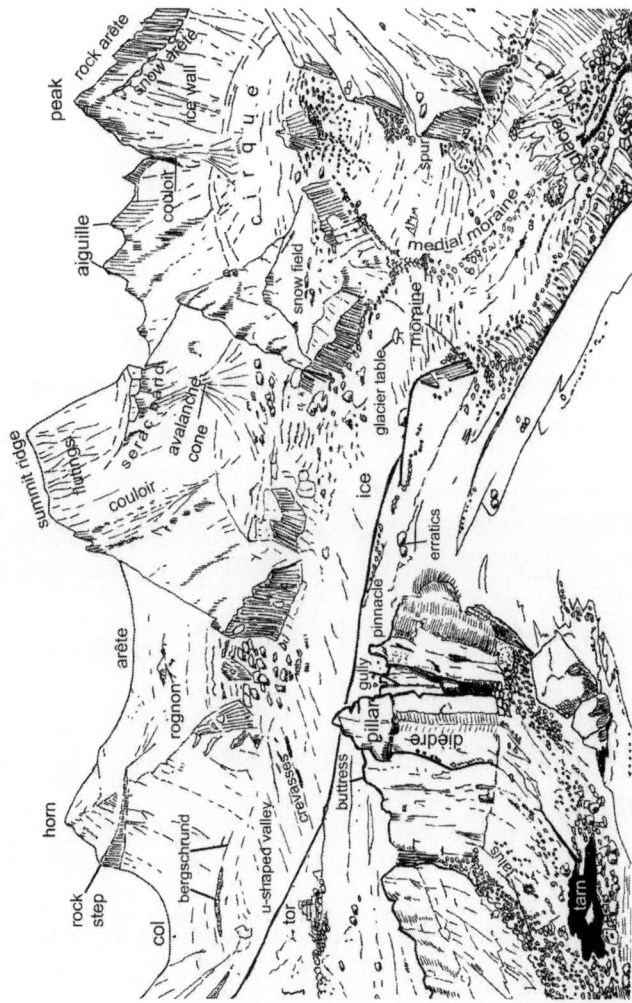

Figure 2.9: A hypothetical mountainscape illustrating many features common to mountains the world over, most of which are, or have been, glaciated. In the left foreground is a cwm (Welsh term) or corrie (Gaelicterm) typical of many found in the British Isles, sculpted by ice which has since disappeared. Beyond to the right is a glacial cirque (French term associated with Alpine conditions). Further left are large snow and ice peaks whose features tend to be more Himalayan (source: Cleare 1979, pages 8 and 9).

3. Undercutting by a river or spring.

Steep slopes are often due to a river undercutting its banks, especially on the outside of curves. The emergence of a spring can have a similar effect **(Figure 2.10)**.

4. Wave erosion.

Waves provide one of the greatest concentrations of energy in nature. The force and direction of their action depends on the depth of water and the configuration of the coastline. As the depth decreases towards the shore, advancing waves are refracted and the crest lines tend to become parallel to the bottom contours. This concentrates waves on to exposed headlands.

Waves, sometimes assisted by tidal currents, erode, transport and deposit materials removed from cliffs and headlands and brought down by rivers. They erode by pounding the coast, abrading it with suspended solids and compressing air into cracks. This is sometimes helped by chemical action, especially against soluble rocks such as limestone. With faulted or jointed rocks the wave attack creates *cliffs*, *caves*, *arches*, and isolated monoliths or pillars rising steeply from the sea called *stacks*.

Waves attack rather like a horizontal saw, cutting a *notch*, and causing overlying rock to collapse. The fallen detritus is removed by *backwash*. The retreat of the cliff leaves a near-horizontal *wave-cut platform* **(Figure 2.11, see also Figure 16.1)**. The most strongly attacked points are headlands and promontories, to which the name *ness* (alternatively *naze*, *naes*, or *nab*) is given (although it may also refer to a spur from a mountain ridge).

Cliffs are cut indiscriminately back into existing landscape features, so that when, for instance, their retreat exposes the cross-profile of a valley, the latter appears to hang above the coast, as in the Seven Sisters on the Sussex coast **(Figure 2.12)**. When a stream erodes such a hanging valley towards the sea, the resulting narrow steep-sided gorge is known as a *chine* in southern England. The detailed form of cliffed coasts reflects contrasts in rock competence at a smaller scale.

Rates of cliff retreat depend on rock hardness, being especially rapid in drift materials. Cliffs in some places such as the Holderness area of Yorkshire have an average loss of 2 metres

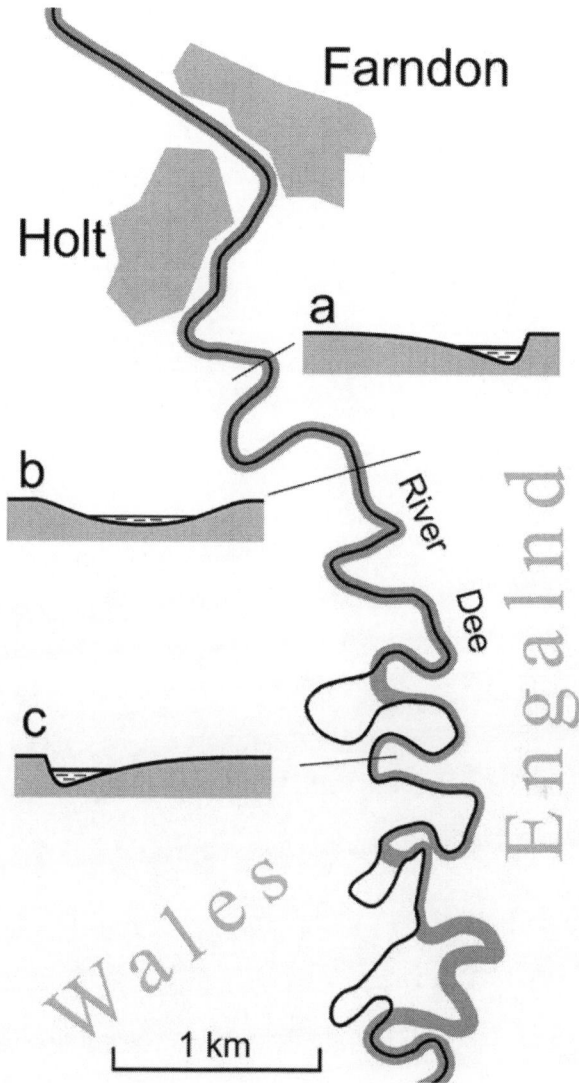

Figure 2.10: Meandering rivers tend to leave a shallow deposit of sediment on the inside of bends and cut a steeper bank on the outside, widening valleys as they do. This map of the north-flowing River Dee shows the profile of its bed at 3 points: a, on a west turning meander; b, between meanders and c, on an east turning meander. The thick grey line is the present course of the river. The fine black line is the border between Wales and England which dates from 1535. When the border was defined it lay along the river course. Shifts that have occurred in the course of the Dee over 5 centuries can thus be seen.

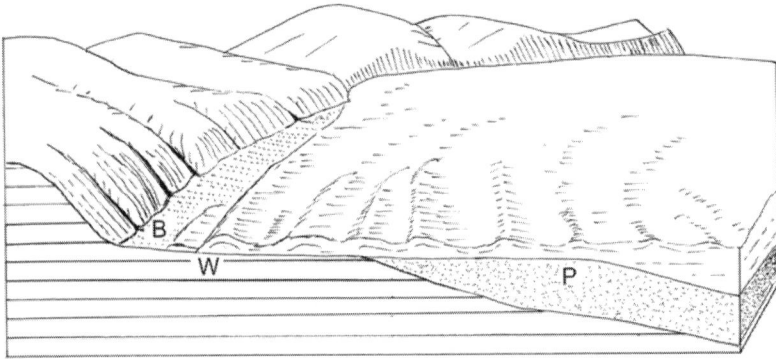

Figure 2.11: A mature coast: (B) beach; (W) wave cut bench; (P) abrasion platform (after Strahler page 528).

Figure 2.12: The Seven Sisters in Sussex from the east showing how the formation of cliffs leaves inland valleys "hanging".

per year, or about 4 kilometres since Roman times. The North Sea storm surge of 1953 removed more than 30 metres of a 2 metres high cliff in eastern England in a single night. In general, however, more land is gained than is lost because the material removed from small retreats of cliffs and headlands is enough to cover relatively wide areas at low elevation. In southern England this accretion has been notable in the Wash, around Romney Marsh and along the extended *spits* (see below) of Spurn Head and Orford Ness.

The British coasts exhibit examples of wave attack on almost every type of rock. There are cliffs in the following materials: volcanic (St Kilda, Skye), plutonic (Land's End), metamorphic (Isle of Lewis), limestone (Portland), sandstone (John O'Groats), and chalk (Beachy Head).

The height of cliffs reflects the erosive force of the waves and the duration over which the land has been exposed to them. High cliffs suggest long continued exposure to stormy seas. Where rock types vary along a coastline, there is an alternation of headlands and bays which reflects their varying competence. This is well shown on a geological map of the southeast coast of England from the Isle of Wight to the Thames estuary, where headlands are invariably on limestone, sandstone, or chalk and embayments on clays or shales **(Figure 2.8)**.

5. Differential erosion of rocks of strongly contrasting competence.

This is especially conspicuous when the hard core of a volcanic vent is exposed by the erosion of the softer surrounding materials, leaving a near-cylindrical *neck* (also called a *plug*) **(Figure 3.3A)**. Well-known examples are the Puy de Dome (France) and the Devil's Tower (Wyoming) **(see Figure 3.10)**.

An oversteepened slope can also occur where a competent rock with a coarse fracture pattern is perched in a way that blocks fall off along vertical fractures. This is commonest where they overlie softer rocks, whose erosion undersaps the overlying strata and determines the rate of cliff retreat. Examples can be seen under *caprocks* especially in arid areas and under cliff faces along the volcanic Whin Sill (Northumberland) **(Figure 2.13)** and wherever hard cap rock protects underlying beds from erosion.

Moderate slopes

Normal erosion causes rocks to *recline* to a gradient which is stable for the structure and material under the particular ambient climate. Once this stable gradient is achieved, they

Figure 2.13: Whin Sill, Northumberland: a basalt dike exposed at the surface because of its greater resistance to erosion than the surrounding Carboniferous limestone.

maintain it relatively unchanged while they *retreat*. This is called *parallel retreat*. This sequence can be seen along coasts where an upper erosion slope formed by parallel retreat is above a cliff undercut by more recent marine erosion. There are intermediate slopes where abandoned cliffs are gradually reclining. All three types of slope can be seen along coasts where a cliff line has been abandoned by the sea, as has apparently happened progressively from Laugharne to Pendine in Pembrokeshire **(Figure. 2.14).**

Figure 2.14: Coastal slopes in Swanlake Bay, Pembrokeshire, looking east. Cliffs (1) are due to undercutting by marine attack. Gentler upper slopes (2) are the product of fluvial erosion processes. These are no longer being undercut, are in the process of reclining towards the gentler fluvially-governed gradient. (3) is the termination of a small alluvial fan deposited by the stream.

CHAPTER 3

Igneous rocks

Rocks and Structure

Igneous rocks form from the solidification of molten lava. They are *extrusive* when the molten rock has solidified on the surface or *intrusive* when this has been below ground. Both types generally give bold landforms without strong directional orientation. Their exact character depends on their mode of origin, chemical composition, crystal structure, and fracture pattern.

The most obvious source of extrusive rock is from volcanoes. These explode, spatter, or pour out material at the surface where it solidifies rapidly **(Figure 3.1).** This gives little time to develop a crystal structure. Without this they are called amorphous. Lava that has flowed over the surface before solidifying is *basalt*. Material blown out in separate particles which has then solidified as an agglomeration of lumps is *tuff*.

Plutonic rocks, by contrast, have solidified from molten masses of lava below the surface. They are generally only exposed at the surface when the overlying layers have been eroded. Because they solidified at depth they cooled slowly, leading to a marked crystal structure **(Figure 3.2)**. The slowest materials to cool form the largest crystals, giving rise to pegmatites when they solidify. Joints between the crystals are widely spaced and this makes the rocks more competent.

Intermediate between these two types are *hypabyssal* rocks. These occur where lava percolates into surrounding rock but does not reach the surface. They are usually long narrow intrusions where lava has squeezed into cracks. This causes an intermediate rate of cooling and a mixed crystal structure with some large and some fine components. A particularly common

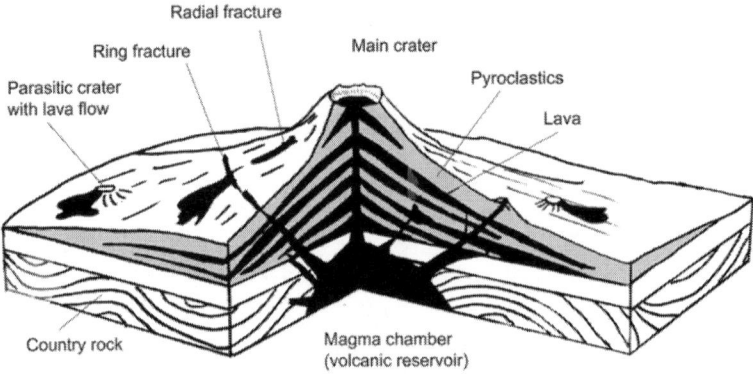

Figure 3.1: Volcano and associated features.

Figure 3.2: The texture of an igneous rock. The material segregates into crystals of various minerals as it cools. These have interlocking irregular or straight boundaries. Crystal sizes range from several centimetres to sub-microscopic. Sometimes crystallization is inhibited and a natural glass is formed (eg. obsidian) (after: Miller p42).

form of hypabyssal rock is *porphyritic* in which large crystals are surrounded by a fine-grained groundmass. This form of solidification is especially common in rocks rich in silica and alumina.

The competence of igneous rocks increases with their chemical acidity, which is directly dependent on their silica content, and is usually indicated by the proportion of quartz they contain. This mineral is light in colour and highly erosion-resistant. Basic rocks have little or no quartz and a higher pro-portion of darker ferro-magnesian minerals such as amphibole, pyroxene, hornblende, and biotite mica. These weather more easily. The aluminosilicate feldspars are intermediate. Of the latter, orthoclase (potassic) and sodium-rich plagioclase are richer in silica than calcium-rich plagioclase, and so are more competent.

Derived soils reflect these differences. Acid rocks yield sandy and gravelly detritus which makes generally poor soils bearing lime-avoiding *calcifuge* vegetation such as heather and bracken. Basic rocks yield soils richer in clay and so more fertile.

Joint pattern is another guide in distinguishing igneous landforms from each other. Rock competence decreases as joints become more closely spaced. Since closeness of spacing increases with the rate of cooling of the rock, volcanic rocks tend to crack more and to be less competent than plutonic.

The combination of mineralogy and fracturing means that rock competence tends to decrease from top left to bottom right of Table 3.1.

The other factors determining rock competence are the width and continuity of joints and their orientation in rela-tion to the alignment of the topography. The narrower and less continuous the joints, the stronger the rock. Also the more the joints are inclined into the hillside rather than parallel to it, the less easy it is for detached detritus to fall, and so the more ero-sion-resistant the rock.

Terrain on volcanic rocks

Solid volcanic ejecta form *cinder cones*, *ash cones*, or *scoria cones*. *Scoria*, also called *volcanic slag*, is dark vesicular material, partly

Table 3.1 Classification Of Igneous Rocks By Chemistry And Crystal Structure			
	Acidic	**Medium**	**Basic**
Crystalline	Granite	Diorite	Gabbro
Porphyritic	Rhyolite	Andesite	Porphyritic basalt
Aphanitic*	Felsite	Felsite	Basalt
Glassy	Obsidian	Obsidian	Tachylite
***** having crystals too small to be seen by the naked eye			

glassy and partly cindery, which is formed by the rapid cooling of lava blisters and bubbles which have been fractured by volcanic eruption.

Ejected volcanic fragments, also called *pyroclastic materials* ('broken by fire') or *tephra*, are classified as shown in Table 3.2.

If the material is ejected forcefully in liquid form by gas pressure it forms a *spatter cone*, composed of agglutinated clots.

Solidified underground reservoirs of igneous material (usually granite) which outcrop at the surface have different forms. *Batholiths* have no observable bottom to the structure and can appear at the surface as outcrops called *plutons*. These can be single or else multiple when linked at depth, as in the granite domes of Dartmoor, Bodmin, and St Austell moors in southwest England. A *laccolith* is an intrusive mass with a pipe-like feeder which forces the overlying sedimentary rocks up into an arch. A *lopolith* is the reverse - a saucer-shaped igneous

Table 3.2 Types Of Solid Volcanic Ejecta		
Size	**Shape**	**Name**
>32mm	angular	blocks
>32mm	rounded or ellipsoidal	bombs
32-4mm	lapilli (little stones) or cinders, scoria	
4-0.25mm	ash	
<0.25mm	fine ash, dust	

intrusion, contrasting with a *phacolith* whose surface outcrop is concordant with the surrounding structures. Both tend to be accordant with the surrounding structure.

Most volcanic material flows out as a liquid. The resulting landforms depend to some extent on the chemistry of the magma. Basic lavas, being richer in iron, are non-viscous, emerge first, and spread longer distances from the source. Acidic lavas are more viscous, emerge last, and solidify more rapidly as short flows or tongues on the slopes of the volcano. Therefore, the central core and the proximal zone around a volcano tends to be composed of more competent acidic rocks which yield bolder landforms and poor sandy soils while the distal zone yields less competent basic rocks but richer soils.

Volcanoes take a number of forms (**Figure 3.3**). *Cratered volcanoes* contain a central *crater*, which when large is called a *caldera*. Around and within a caldera are usually several lava vents with smaller volcanoes known as *adventitious* or *parasitic*

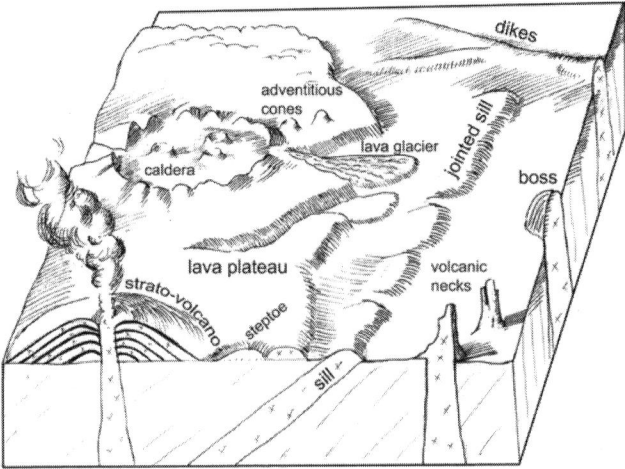

A

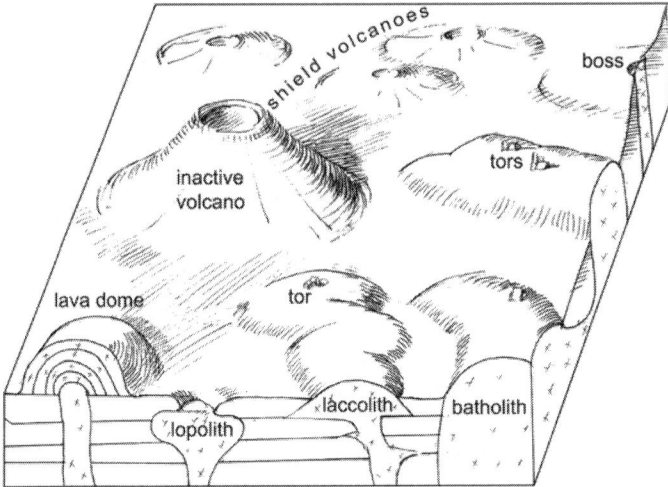

B

Figure 3.3: Diagrams of possible relations between igneous rocks and relief. Igneous rocks are usually the more competent and thus the more prominent features of eroded landscapes. The different patterns of penetration of magma into sedimentary rock give recognizably different landscape types (after Sparks page 150, drawings by Julian Mitchell).

cones or *volcanic cones*. A caldera containing such features is called a *composite volcano*.

When the ejected magma is fluid, it either spreads relatively widely around the vent forming a low dome called a *shield volcano*, or else follows a valley as a *lava glacier*. Where the volcano has no crater, it is called a *lava dome*. Where it builds up by numerous flows in the same place it is a *stratovolcano*. If formed entirely of layers of solid ejecta it is called *simple*; if these alternate with former lava flows, it is called *composite*. Where a lava flow completely surrounds an isolated hill or tract of land, the latter is called a *steptoe*. Erosion sometimes reduces volcanoes to residual cones around a central pinnacle. These are called *volcanic necks*.

Lava often emerges along deep cracks in the earth's crust forming a chain of volcanoes. The Hawaian islands are an outstanding example of this. They have been formed in chronological series as the northwesterly movement of the Pacific Plate has carried a succession of volcanoes away from a stationary mantle plume (jet of partially molten rock) located at the southern tip of the island of Hawaii. As a result, the volcanoes show increasingly weathered and eroded forms from southeast to northwest along the island chain.

Freshly ejected lava assumes two main types of surface. It sometimes traps gases under its solidifying crust. These accumulate in pockets and erupt, breaking up the crust. This forms the rough fragmented surface known as *aa*. Where the lava solidifies in a corded or rope-like form, it is called *pahahoe* **(Figure 3.4)**. If the crust solidifies but the liquid lava below it drains away, the resulting hollow is called a *lava tunnel*. Occasionally the surface crust will have circular mounds over caves called *monticules*. These are thought to have been caused by trapped gasses which could not escape before the surface solidified hermetically.

Lava can also solidify into a columnar pattern as at the Giant's Causeway (Antrim), Fingal's Cave (Isle of Staffa), and the Sgurr of Eigg **(Figures 3.5 & 3.6)**. This is due to the lava contracting towards each of a series of equally spaced centres as it cools. Since neither the physical conditions nor the rocks are absolutely uniform to ensure perfect symmetry, the actual result is a set of columns with three to eight sides, six being the

Figure 3.4: Rope-like pahoehoe lava which has solidified quickly, Arizona, USA.

Figure 3.5: Columnar basalt seen at the Giants Causeway in Northern Ireland.

commonest number. Cross joints divide the columns into short lengths.

Where the main volcanic eruption is of superheated water or steam through a hole in the ground, it results in a *geyser*. This erupts with regular periodicity due to the rebuilding of water temperature and pressure after it has been relieved by the ejection. "Old Faithful" in Yellowstone Park is an example with great regularity. When the water is charged with minerals such a silica or lime, its evaporation leaves a mound with a central crater at the point of emergence.

When the main eruption is of gases, the result is a narrow funnel or bowl-shaped crater called a *maar*. Its perimeter is made up of brecciated (sharply fragmented) local rock and the crater itself often contains water.

In humid temperate climates, weathering soon modifies volcanic surfaces to give moderately steep erosion slopes with knobs and hollows reflecting local differences in rock competence. Jebel Uweinid, Jordan, shows the protective effect of a capping of lava (**Figure. 3.7**)

In arid and tropical climates gullies, called *barrancas*, develop radially around volcanoes. More advanced dissection isolates triangular or wedge-shaped *planezes* with the narrow

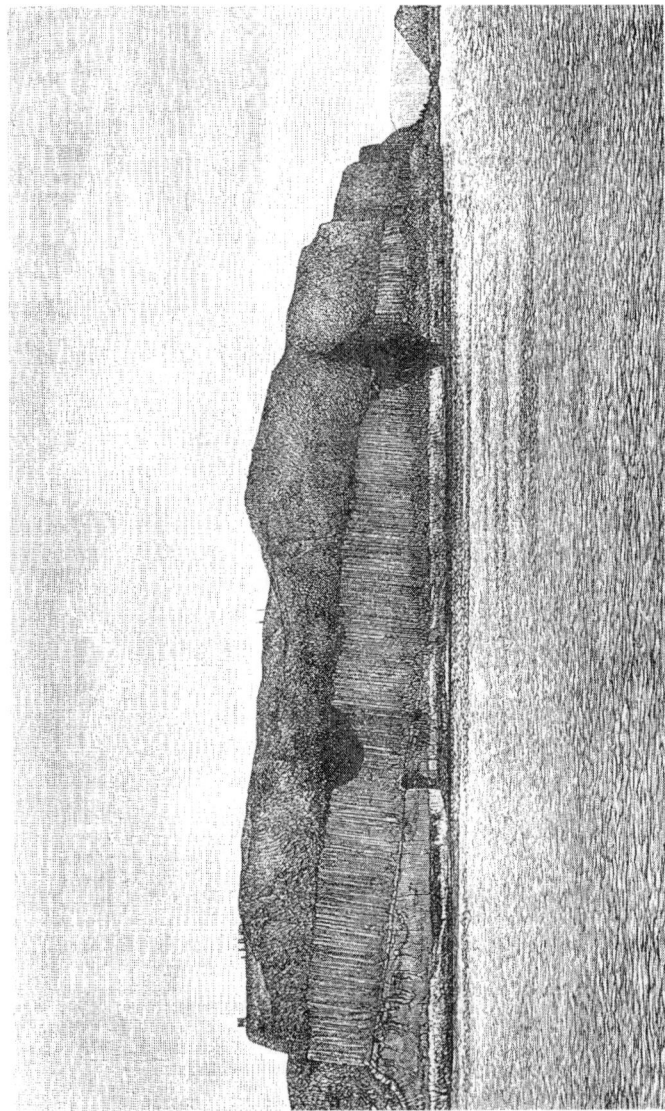

Figure 3.6: View of the south side of the island of Staffa of the west coast of Scotland, showing the bedded and columnar structure of the basalt. The rock in which the cave on the left has been eroded is a volcanic tuff underlying the basalt. Fingal's cave is on the right. The caverns testify to the enormous erosive power of the Atlantic breakers (source: Geikie (1887) Figure 48).

Figure 3.7: Eroded remnant of a lava .ow capping other rocks: Jebel Uweinid, Jordan.

ends pointing up-slope. Overflowing crater lakes can cause a mudflow of volcanic ash and water, known as a *lahar*. The complete removal of the weathered detritus around a volcano may expose its central core as a neck or plug. When erosion leaves a landscape dominated by such hills, as in parts of Mali and Chad, it can be called *polyconvex*; where it leaves a landscape of tablelands edged by concave slopes, it can be called *polyconcave*.

Terrain on plutonic rocks

Plutonic rocks have larger crystals than volcanic rocks. The first and smallest crystals to segregate are chiefly the ferromagnesian minerals. These, because of their higher specific gravity, sink to the bottom of the magma chamber, leaving the more siliceous and volatile materials towards the top.

As with volcanic materials, the competence of plutonic rocks in the landscape depends largely on their quartz content and the coarseness of fracture spacing. This is well illustrated in the Scottish Highlands where the harder and more quartzose the rocks and the more regular their system of joints, the loftier and more rugged their heights.

The fracture pattern in plutonic rocks is usually rectilinear with most fracture planes either parallel to or perpendicular to

Figure 3.8: Onion-skin weathering on a granite slope in central Sudan.

outer surface of the rock. Fractures also increase in frequency towards the surface because this is the locus of the most rapid cooling and is the area most affected by any release of compressive pressures when overlying rock is removed. Where the joints parallel to the surface are dominant and closely spaced, we find a type of weathering known as *exfoliation*, also called *desquamation* or colloquially as *onion-skin weathering*. Where it shapes a whole landform, it creates an *exfoliation dome* (**Figure 3.8**).

Slope steepness reflects these controlling factors. The wider spaced the jointing, the larger the spalls and the steeper the slope. Steepness increases still further if the detached spalls crumble quickly because this prevents them from accumulating to provide an erosion-resistant mantle on the footslope. In general, the wider the ratio between spalling size and the size into which the spalls crumble, the steeper the slope.

We know more about granites than about other plutonic rocks because they tend to be the most widespread hill formers. They occur widely in the relatively level hard-rock shield areas which cover much of Scotland, Scandinavia, North America and Africa. In humid temperate climates granite tends to form uplands with relatively uniform slope angles and with a pronounced parallelism of ridges and valleys and with detached sub-circular boulders (**Figure 3.9**). Summits tend to be rounded and slopes linear, as in mountain areas such around Ben Nevis and even on lower undulating hill country such as Dartmoor and Bodmin Moor. The Isle of Skye contains a spectacular illustration of the effects of chemical differences. The smooth rounded contours of the granite Red Hills contrast with the rugged outlines, many sharp peaks, and extensive rock wastage of the Cuillins on more basic volcanic material called *gabbro*.

In tropical and arid countries, plutonic rocks often form isolated hills rising abruptly from the plain. These are called *inselbergs*, or if formed on granites, *bornhardts* (**Figure 3.10**). Their slopes have relatively uniform gradients but are generally steeper than in temperate climates because of the more intense weathering and absence of vegetation. They sometimes have footslopes covered with fallen rock fragments called *talus*. Smaller rock domes with near-vertical sides and a sugarloaf shape are variously called *stocks* in Europe or *kopjes* in South Africa.

Granite uplands in arid and semi-arid areas commonly show a sharp break of slope at the hill foot where a *nick point* separates the upper hillslope from a gentler *pediment* below which is covered by a scatter of detrital material (**Figure 3.11**). The angular contrast between the upper and lower slopes and the resulting sharpness of the nick point are especially marked in granitic rocks and schists. A lateral coalescence of pediments is known as a *pediplain*.

It is believed that pediments are formed either a) by the parallel retreat of the mountain front with the nick being caused by a change of phase when torrential runoff down the mountainside becomes arrested after accumulating a critical sediment load, b) by lateral corrasion by streams which anastomose after debouching from the mountain front, or c) by weathering under an earth mantle followed by exhumation. None however, of these explanations is satisfactory, as no pediments are observed to be forming today. Although all these processes operate in different proportions in different areas, it has been suggested that pediments are mainly erosional features caused by deep currents during the recessive stage of a submerging flood (Oard, 2004).

Figure 3.9: Erosion slopes on granite in semi-arid Arizona. They are characterized by a uniform slope angle and surface weathering into boulders. Because sharp edges and points are most exposed to such weathering they tend towards rounded forms.

Figure 3.10: Granite bornhardt with talus slopes: Devil's Tower near Black Hills, Wyoming (after Lobeck page 647).

Terrain on hypabyssal rocks

Terrain on hypabyssal rocks generally reflects the difference in competence between the intrusions and the rocks into which they are intruded. Where the intrusions cut across the existing rock structures they are called *dykes* (**Figure 3.12**), where they follow bedding planes, *sills* (see **Figure 3.3**).

Dykes may radiate from, or be in concentric rings around, the magma source. Radial dykes may be in swarms, as exemplified by those which extend from a focal area in the Isle of Mull across southern Scotland and into northern England.

Concentric dykes are less common. They may dip inwards or outwards from the magma centre. They dip inwards where upward magmatic pressure has caused near-circular fractures in the form of inverted cones which, when filled with magma, form *cone-sheets*. These are often multiple

Sharp junction of slope angles

A

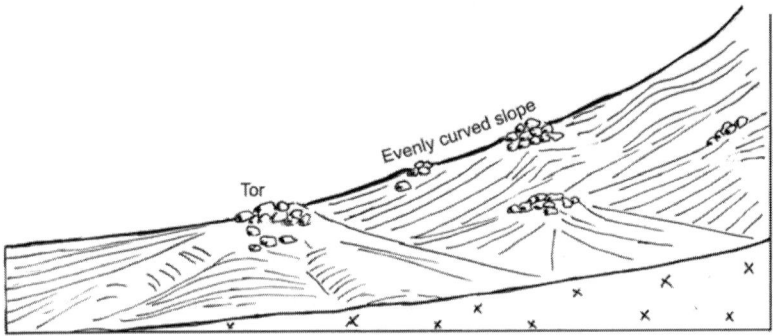

Evenly curved slope

Tor

B

Figure 3.11: The angular junction between upper slopes of granitic mountains and the pediments at their bases (heavy line) (A) contrasts with the curved junction in non-granitic mountains (B). Views from Mohave Desert, California (after King and Schumm page 42).

Figure 3.12: Basalt dykes in limestone, Abdul Kuri Island, Indian Ocean.

and are well displayed on Mull, Ardnamurchan, and Skye in Scotland. Dykes which dip outwards or are vertical are called *ring-dykes*. They occur where the rocks above the magma source have been unsupported and break along circular fractures which then fill with magma. They are much thicker than cone-sheets.

Dykes, sills, and cone-sheets are often composed of more competent materials than their surroundings so that they are left upstanding when these are eroded away. There are examples of dykes standing as walls radiating from a volcanic neck in Shiprock, New Mexico, and of ring-dykes in the Red Sea Hills of the Sudan. Since sills follow the stratification of surrounding rocks they are sometimes isolated as cuestas as can be seen in Whin Sill across the north of England.

Small features

Granitic areas are often recognizable from the frequency of residual quasi-rounded bosses which give a knobbly appearance when seen from a distance (**Figure 3.13**). Where these are large and separate they are called *tors*; where less so, *nubbins*. Both can also occur on other types of rock, but are most characteristic of granites. Both forms are often accompanied by *perched blocks* (**Figure 3.14**). Sometimes the uppermost block of a tor-like feature is supported on a narrow stem resembling a mushroom. These are called *mushroom rocks* or *cheesewrings* after an example on Bodmin Moor (**Figure 3.14**). Tors on Dartmoor are surrounded by fields of small blocks called *clitter*, and coarse-grained weathered debris called *growan*. Clitter is sometimes not scattered randomly but in lines and lobes downslope. These mark the former pathways of the sludging process, a solifluction in which the blocks are moved downhill over the frozen ground beneath. More generally, accumulations of decomposed granite are called *grus*.

Tors are thought to originate where the granite lies below the land surface. Parts of the rock mass with widely spaced joints are preserved more than their more closely jointed surroundings. When they are exhumed (exposed to air) above the surface by circumdenudation, the larger unweathered rock cores remain upstanding as isolated hillocks.

Figure 3.13: Weathering of a granite tor along fracture lines near the top of Bennachie, Aberdeenshire (Source: Geikie (1887) page 17).

Some smaller features can be found on a variety of igneous rocks of all kinds. Step-like benches, known as *Piedmonttreppen*, can occur around the flanks of mountains. They were first recognized in the Blackforest by the German geologist Walther Penck, who thought that they resulted from episodic uplift of an expanding dome, the summit area of which would be the "Primarumpf".

The foot of slopes or even of large boulders may be steepened by *flaring*, probably due to accelerated weathering by a greater persistence of moisture at the hill foot. There may also

Figure 3.14: Boulders of disintegration resulting from weathering of jointed granite. The Cheesewring - Bodmin Moor.

be small vertical depressions at the foot of scarps which appear to be due to increased weathering in this zone because of more persistently moist conditions.

Small surface irregularities are characteristic of granite and other plutonic rocks mainly in tropical and subtropical areas. Some are called *pseudo-karst* because of their resemblance to solution features on limestone, but they are in general smaller and devoid of its characteristic hydrological flow patterns. They include:

Kluftkarren (French *cannelures*) are small U-shaped groves and flutings on the rock surface due to weathering by water flowing down its sides. They tend to be most frequent and narrow on slopes of over 50 degrees, especially between 50 and 70 degrees. They are fewer but wider between 30 and 50 degrees, and do not occur on slopes gentler than 30 degrees.

Tafoni in igneous rocks are shallow caverns or hollows partially enclosed by *visors*. They are initiated by soil moisture attack and developed by salt crystallization and temperature variations. The external side of the visor is case-hardened usually by chemical impregnations with silica or iron.

Gnamma holes. This is an Australian term for rock holes which can be up to 2 metres deep and 3 metres wide, often formed by the accelerated weathering of gaps in a quasi-horizontal rock surface which has been case-hardened in a similar way to tafoni.

Splits and cracks are common in igneous rocks. They are usually due to the exploitation of surface fractures by frost. They sometimes have polygonal patterns which may have originated by weathering the rock surface while still covered with detritus. On exposure to air tangential expansion can cause this to develop breadcrust-type polygonal cracking.

CHAPTER 4

Sedimentary rocks

The earth was probably originally molten, solidifying into igneous rocks from which all land forms ultimately derive. Weathering fragments the rock surfaces and detaches particles. Erosion then carries them away and deposits them as sediments. Compression, heat, and binding agents over time change them into solid masses of sedimentary rock. The same processes then repeat the sequence, creating a new generation of sediments (**Figure 4.1**). This gives layered beds showing depositional stages from bottom to top. Today sedimentary rocks cover about 75 percent in the earth's land surface.

Sediments are normally classified by their mode of formation into mechanical, chemical or organic types. Mechanical sediments are deposited by water, ice, or wind. They are composed of stones, sand, silt, and clay which have solidified into conglomerates, sandstones, siltstones, shales, and certain limestones. Chemical sediments are much less widespread. They are precipitated from water solution either by evaporation or chemical change. The commonest types are calcite, gypsum, and rock salt. Organic sediments are the products of plant or animal organisms which are deposited in water or air. They include coral, shell limestones, chalk, and coal.

Cyclical changes in sea level can give a sedimentary sequence alternating between some or all of these types. In the Ice Age the alternate freezing and thawing of polar and mountain ice caps caused such changes. Falling sea levels resulting from ice advance led to more erosion and accelerated deposition of coarser materials. Rising sea levels led to the deposition of finer materials in deltas and estuaries, and the enclosure of lagoons which accumulated organic materials of both plant and animal

Figure 4.1: A tabular landscape of horizontal sediments: the plateau and canyons of the Colorado (Source: Geikie (1903) frontispiece).

origin. Where the sea level subsequently fell back again such deposits were exposed to movement by wind. Where a single cycle of sea level change causes such depositional changes, the resulting deposit is called a *cyclothem*. Normally the sequence begins in fresh water which then becomes increasingly brackish until it is sea water, finally returning in the same sequence. It represents an oscillation of conditions between terrestrial and marine.

Sedimentary Plateau Landscapes

Sedimentary rocks are usually covered with looser surface layers of soil and vegetation than are igneous rocks. We can only see them when some geological action has raised them above the surface or erosion has cut into them. The simplest case is when horizontally bedded sediments are uplifted into a plateau without tilting, giving rise to a *tabular* landscape. The harder bands form the caprocks and steps in valleys. The softer materials are the "butter in the sandwich". *Inliers* occur where an enclave of underlying older rock occurs at the surface. When a plateau formed of alternating hard and soft beds is deeply incised by valleys and canyons, the resulting landscape of eroded slopes on softer beds between fairly level terraces on the harder beds is called *storeyed*. Plateau surfaces tend to be level and stony with thin soils. In arid areas like North Africa the wind removes fine particles leaving a stony plain called a *hamada* (**Figure 4.2**).

The edges of plateaus have escarpments with four recognizable sections (**Figure 4.3**). Highest is the exposed *scarp face*. Below this is the *debris slope* where materials move down mainly by gravity, then the *footslope* where they are mainly carried down by running water, and finally the *toeslope* where they are carried off in a different direction i.e. along the valley towards the sea rather than down the slope. Viewed from above, the scarp slope shows *promontories* and *reentrants*. The latter are the points where most rainwater reaches the dipslope and so the edge backwears most quickly. The slopes below the promontories backwear slowly because they receive the least runoff. The progress of erosion therefore tends continually to accentuate the contrast between the promontories and reentrants. The

Figure 4.2: Hamada in Jordan.

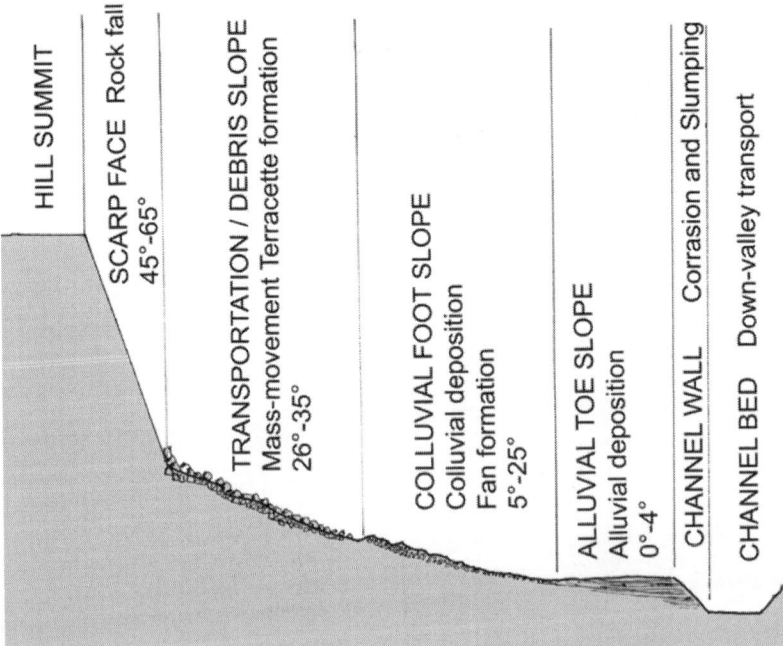

Figure 4.3: Slope units on a hypothetical land surface.

side slopes between the two are intermediate and backwear at intermediate speed.

The effect of this process is to multiply the number and round the ends of promontories at the expense of other portions of the plateau edge. It culminates in a proliferation of small quasi-circular outliers. When dissected by erosion, this develops an irregular margin and detached *outliers* (**Figure 4.4**). Such outliers are common enough to receive the name *mesa* in North America, *tafeltop* in South Africa, and *gara* (plural *gour*) in North Africa. When the caprock is finally eroded away to leave a conical summit, the hills are called *buttes* in North America, *spitzkops* in South Africa and *tent hills* in Australia. A partially developed example can be seen in England at Wittenham clumps near Oxford, an outlier of the Lower Chalk.

The reason that the plateau edge does not retreat evenly is that the rate of erosion increases with the area of catchment from which the water comes and this increases with the size of the plateau and tends always to run into reentrants rather than over promontories. Both water supply and shape thus tend to multiply insular hills and make them circular. This explains their prevalence in many areas.

In a tabular landscape the least competent strata are usually clays. They are impermeable and stop water soaking down through the rock. It must therefore run sideways out of the rock and emerge as springs at the upper edge of clay outcrops on the valley sides. Clays can also be squeezed outwards

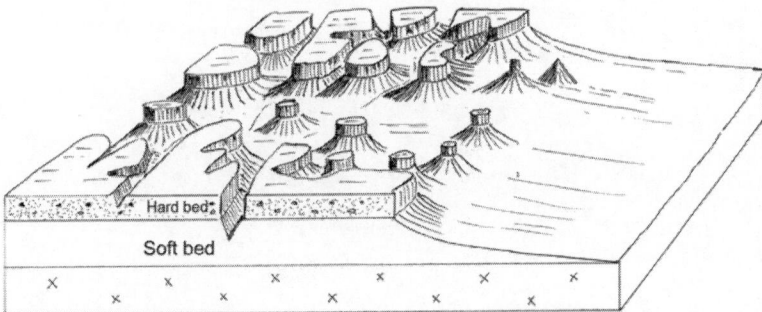

Figure 4.4: The effect of a hard capping bed in saving the underlying material from erosion.

into the valleys by the weight of overlying strata, which then fracture, bend downwards, and drape over the clay outcrops. This is called *cambering*. It can be seen in the Cotswold valleys where the oolitic limestones drape over the valley edges and collapse into landslips because of the plasticity of the underlying Liassic clays. Clays are more vulnerable to erosive removal by water than the strata above and below them. If clay in such a position is washed out, overlying strata can bend downwards in a *knee-fold bend*. Erosive removal of soft beds between harder ones can also lead to the formation of *caves* and *natural arches*. This especially occurs where limestone overlies an impermeable bed (**Figure 4.5**). The cave interiors can show a characteristic range of features. These often include *spongework*: patterns of shallow depressions with intervening rim-like partitions on the walls and ceilings, and a covering of detritus on the cave floor.

Tilted Sedimentary Landscapes

When sedimentary rocks are tilted, the resulting landscape is aligned in the direction of the tilt (**Figure 4.6**). The slope angle down the strata is the dip, the compass direction of the aligned landforms is the strike. The scarp faces of resulting cuestas overlook *vales* as for example the Chiltern Hills overlook the

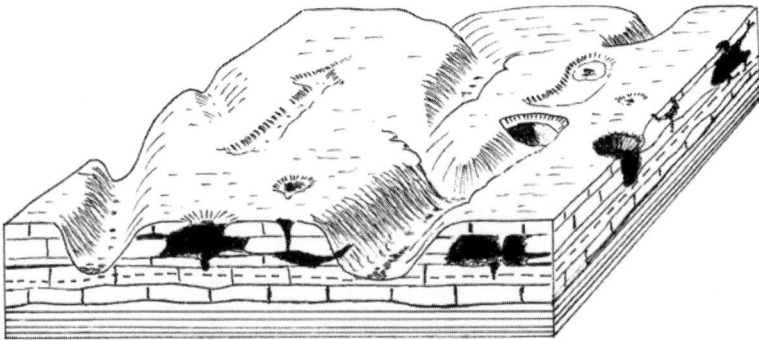

Figure 4.5: Karst topography showing caverns, sinks, solution valleys and natural bridges. The dashed line represents water table. The area shown is about 1/4 square mile (c2/3km^2) (after Longwell, Knopf and Flint 1941 p112).

Figure 4.6: Tilted sandstone on beach near Port Elizabeth, South Africa, viewed in the direction of strike with dip angle from top right to bottom left.

Vale of White Horse and the Cotswold Hills overlook the Vale of Severn in southern England. Eroded anticlines give cuestas with escarpments facing each other as in the Weald **(Figure 4.7)**, while synclines give escarpments facing away from each other as in the limestone hills of the Paris Basin around the French capital.

Erosion attacks the exposed edge of the cuesta and streams develop down both the scarp face and the dipslope. Scarp-face streams wear back and can form outliers. The steeper the angle of dip the more regular the escarpment and the fewer the outliers. Where the landscape includes an alternation of hard and soft beds, it takes on a uniclinal *cuesta-and-vale* character oriented in the direction of geological *strike*. The stronger and thicker the cuesta-making strata, the higher and bolder are the cuestas. The weaker and thicker the intermediate strata, the broader, lower, and smoother are the inter-cuesta vales. The most steeply dipping strata can form straight narrow ridges, sometimes called *hogbacks* after the example located between Guildford and Farnham in Surrey.

When steeply tilted strata are isolated into triangular shapes pointing upslope, they are called *flatirons* if the strata are thick and *chevrons* if they are thin (**Figure 4.8**). When the strata are vertical and a competent bed stands proud of the rest, it called a *rock wall*.

The relative height of neighbouring cuestas affects the position of subsequent strike streams. These are usually driven over towards the smaller of their bounding cuestas because of the greater amount of detritus coming from the larger ones. An example can be seen west of Oxford where the upper Thames is driven southwards against the Corallian cuesta by the greater mass of the Cotswolds, and the Ock is driven northwards against it by the greater mass of the Chilterns.

In cuesta-and-vale country, also called scarp-and-vale country, the drainage system tends to be of the type called *trellised*. *Strike streams* follow the vales. *Dip streams* drain down the dipslopes, *anti-dip streams* down the scarp slopes. Where a dip-stream and an anti-dip stream each cut backwards and meet they form a gap called a *recession col*, sometimes called a *wind-gap* (**Figure 4.9**).

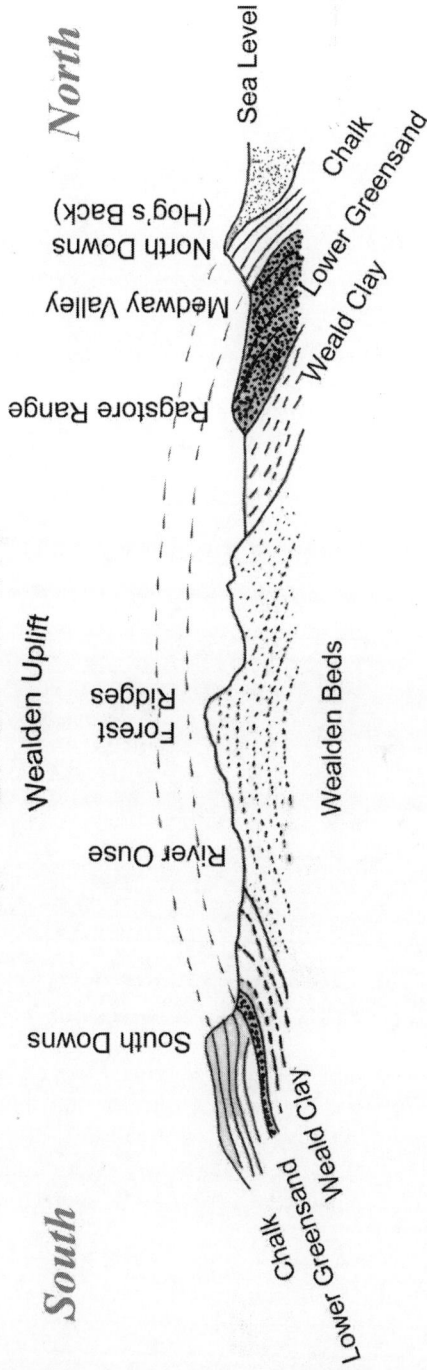

Figure 4.7: Section through the Wealden uplift in Kent and Sussex (after Mackinder page 88).

South

North

Sea Level

Chalk

Lower Greensand

Weald Clay

North Downs (Hog's Back)

Medway Valley

Ragstore Range

Wealden Uplift

Forest Ridges

Wealden Beds

River Ouse

South Downs

Chalk

Lower Greensand

Weald Clay

Figure 4.8: Flatirons (large) and chevrons (small) in sedimentary rocks (after Lobeck page 534).

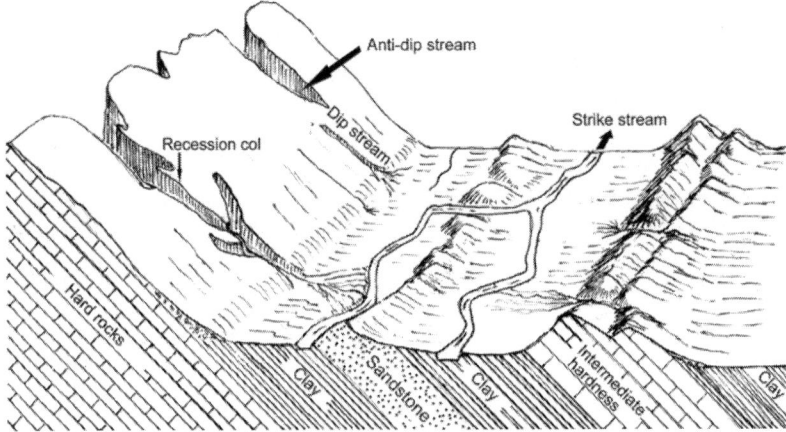

Figure 4.9: Cuesta-and-vale country showing how trellised drainage has followed rock type and structure. Note how the limestone (brick pattern) stands up partly because of its permeability and partly because of its hardness. Sandstone is here shown as intermediate, though is sometimes hard, and clay occupies the lowest ground, attracting the outflow drainage lines.

Where a stream cuts through a cuesta, it usually means that it antedates it and was able to maintain its transverse course as the cuesta emerged. Such a stream is called *antecedent*. Where its direction still follows the slope of the land on which it originated, it is called *consequent*. Streams which have developed along faults or vales as tributaries to the consequent are called *subsequents*. These receive *obsequent* streams draining the scarp slopes, *resequent* streams draining the dipslopes.

Coastal landforms reflect the inclinations of strata. Horizontally bedded rocks give relatively even coastlines. Tilted sediments have a pattern of parallel inlets separated by peninsulas, often terminating in islands. These coastal features depend on the angles of strike and dip of the strata in relation to the direction of the coastline, and on rock hardness. Harder rocks can form caves, windows, natural bridges, and offshore stacks, especially where the bedding is horizontal or vertical. The resulting cliffs are most stable when the rocks dip inland and least stable when they dip seawards.

Wind charged with fine particles scours sedimentary rocks into streamlined landforms. When these are small they are called *yardangs*. *Zeugen* are the special case where the wind has abraded furrows in soft layers under a caprock. On a wider scale, as in parts of the Sahara, the country can be eroded into a pattern of evenly spaced parallel ridges and furrows, which occasionally include hollowed-out *windows* and even *arches*.

CHAPTER 5

Metamorphic rocks

Metamorphic rocks are igneous or sedimentary rocks which intense pressures, heat and the intrusion of other materials have radically changed. Some call their appearance "crystalline slaty" (**Figures 5.1 & 5.2**).

They generate landscapes with broadly the same types of slope features as igneous rocks, although with a greater directionality of fault lines, ridges, and valleys called *lineation* (**Figure 5.3**). Their competence increases with their quartz content and with the spacing of joints. In arid and semiarid areas they show the same sequence of hillslope-knickpoint-pediment as do igneous rocks of similar mineralogical composition.

Intense faulting and folding of sediments can lead to crumpling and overthrusting (**Figure 5.4**). The resulting land surfaces are complex and can be further complicated by the isolation of the overthrust rock structures, called *klippen*, by erosion of less competent surrounding material. The topography generally is of ridges and valleys reflecting the direction of the main tectonic forces (**Figure 5.5**). Level plateau surfaces are absent.

Metamorphic rocks are especially vulnerable to weathering and erosion where they have been chemically altered. Such alteration often results from *contact metamorphism* at the edges of igneous intrusions which creates a zone of soft kaolin and segregated minerals containing metallic elements. It also occurs where they are weakened by fragmentation into *shatter belts* along faults. The drainage network tends to follow and pick out the lines of weakness thus created.

The commonest types of metamorphic rock are quartzite, gneiss, schist, slate, phyllite, and marble. They are often associ-

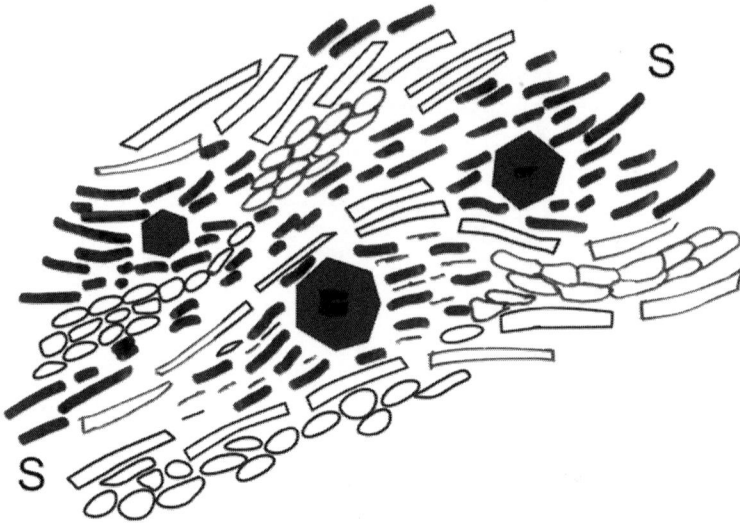

Figure 5.1: The texture of a metamorphic rock - a more or less regular arrangement of mineral fragments and crystals: S-S = plane of cleavage. Approximately natural scale.

ated in the same area. Quartzite (**Figure. 5.5**) is probably the most resistant to erosion of all rock types and is almost always elevated in the landscape. It tends to weather into large rectangular blocks which form talus masses below almost vertical cliffs. Hill summits are generally well rounded and can appear whitish, as for instance in the capping of numerous hills in the Scottish Highlands, including Suilven (Sutherland), Ben Liathach (Ross & Cromarty), Schiehallion (Perthshire) and the Paps of Jura. Gneiss and schist have a pronounced alignment of mineral particles, and are *foliated* (a term based on their similarity to tightly packed leaves). The alternation of ridges on quartzite with valleys on darker more vulnerable minerals can give a corrugated "grain" to the country, with irregularities which can range in width from a less than a metre to more than a kilometre.

Gneiss consists of micaceous layers alternating with granular feldspars and quartz (**Figure. 5.6**). It is almost as resistant to erosion as quartzite, forming a generally rounded upland

Figure 5.2: Gneiss with pegmatite veins, western Scotland (Source: Geikie (1887) page 73).

Figure 5.3: Lineated metamorphic country weathered into aligned mounds picked out by competent basalt dykes along parallel fractures, Abdul Kuri island , Indian Ocean.

topography. Where the land surface has been maturely eroded by ice-scouring, as in the western coastal areas of Sutherland and Ross and in the Outer Hebrides, it gives *knob-and-lochan* country (**Figure. 5.7**).

Schist is an intensely metamorphosed foliated rock with minerals segregated into fine layers usually rich in the mineral mica. It differs from gneiss in having a lower content of feldspars, a more flaky structure, and in being less competent.

Slates are intensely compressed shales with a platy cleavage. Their erosion resistance is less than that of schists and is directly proportional to their quartz content. A local impregnation with silica can form a significant topographic high or coastal headland. Slates tend to form rounded convex-topped hills and moderately steep linear valley sides of which Skiddaw in the Lake District is a good example.

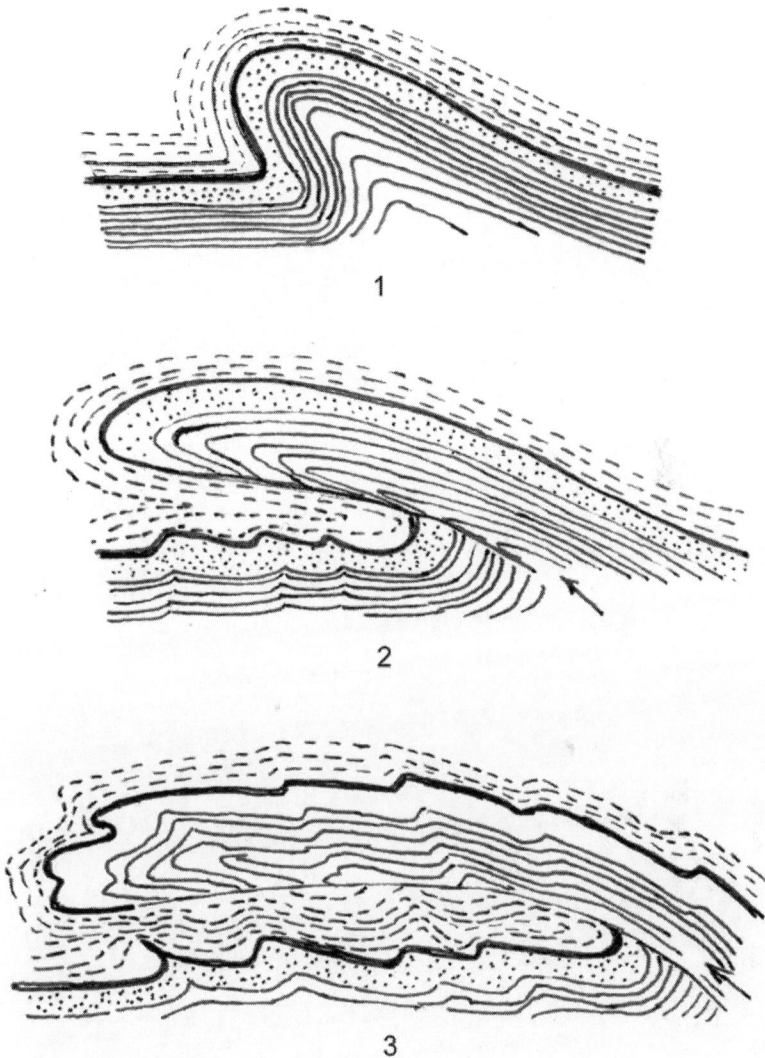

Figure 5.4: Development of a fold nappe by shearing through a re-cumbent fold. Note how lateral pressure creates a fold (1), then causes a fracture (2) so that the right hand limb of the anticline overrides and covers the left hand limb along the fracture line (3) (after Hills 1965 page 55).

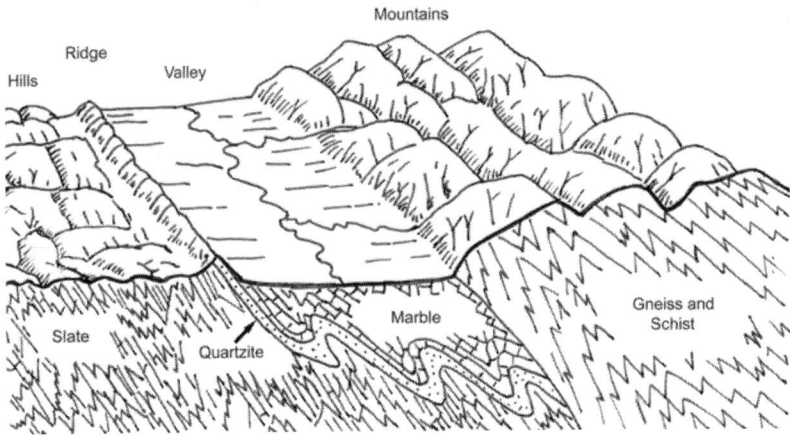

Figure 5.5: Characteristic landforms on maturely eroded metamorphic rocks in a humid climate. Note how gneiss, schist, and quartzite stand bold in the landscape, slate less so, while marble, because of its solubility, is removed and forms the lowest ground. (after Strahler (1965) page 373).

Figure 5.6: Gneiss in New Hampshire. Note the foliation in the rock due to the intense pressures and heat during its formation causing minerals to align themselves in the direction of stress.

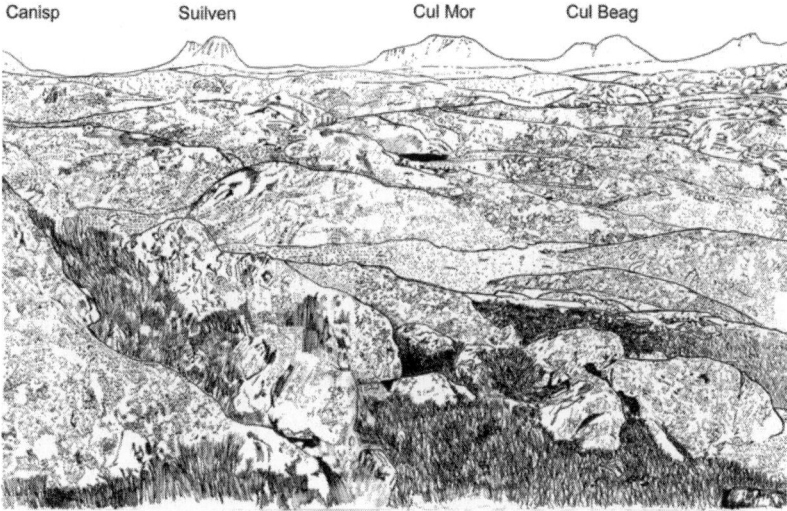

Figure 5.7: Knob-and-lochan country on gneiss just north of Stoer, Sutherland, Scotland. Note the directionality imposed by lineation in the gneiss from right foreground to left middle ground. Mountains in the background are on overlying Torridonian sandstone (from a photograph).

Phyllites are intermediate between schists and slates, finer grained than the former but less perfectly cleaved than the latter, and they therefore give rise to intermediate topographic forms.

Marble is compressed and crystallized calcium carbonate and is distinguished from other forms of the same material by its density and ability to take a polish. Pure marble is white but iron and copper impurities give pink or green colours respectively. Marble gives landforms like limestone except that its cleavage, rather than being rectilinear, is often broken into large blocks by joints. Owing to its solubility it succumbs readily to weathering in humid regions, but may survive destruction better than the usually more resistant igneous and sedimentary rocks in arid regions.

CHAPTER 6

Siliceous sediments

Siliceous sediments are composed of quartz particles in a matrix. Landscapes developed on them reflect their physical composition and fracture patterns. The commonest rock type is *sandstone* laid down as a sediment by running water or wind **(Figure 6.1)**. When its component particles are sharp rather than rounded, the rock is called *gritstone*; when they are fine, *siltstone*; when it includes stones, *conglomerate*; when rich in feldspar, *arkose*; when rich in glauconite *greensand*; and when rich in clay, *greywacke*, now more commonly called *turbidite*. *Ferruginous sandstones* are cemented by iron oxides and are usually brown, red or yellow. Sandstone grains can be cemented together by impregnation with silica, known as *silicification*. Under heat and pressure this leads to crystallization as quartzite. The movement and coalescence of silica in calcareous rocks can form *flint* nodules.

Where sandstone beds vary in erosion-resistance the harder ones protect those beneath them. Suilven **(Figure 6.2)**, one of the most dramatic and prominent peaks in Scotland, stands up because its capping of quartzite has protected the underlying softer sandstones from the erosion which has removed them all around.

Sarsens are an example of fragments of cemented sandstone. These are boulders thought to result from the patchy and irregular hardening of sandy materials by silica. In Britain, they are commonest in the chalklands of Berkshire and Wiltshire, where they are thought to be residues of the sandy Reading Beds which formerly covered the Chalk, after the removal of the softer uncemented sands by erosion. They form a conspicuous scattering of boulders over otherwise smooth

Figure 6.1: Wealden sandstone at Eridge Rocks, Kent. Note the horizontal fractures following the layers of the strata and the rounding of weathered sandstone exposures because of the preferential removal of prominent points and edges.

A

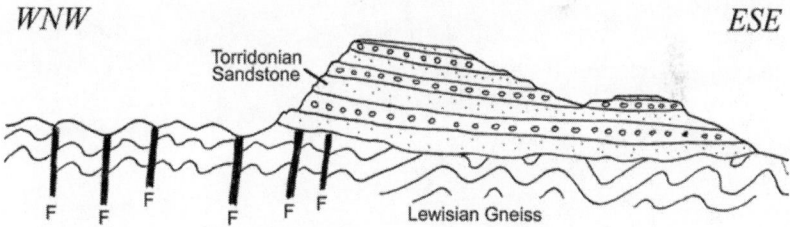

WNW　　　　　　　　　　　　　　　　　　　　　*ESE*

Torridonian
Sandstone

F　F　F　　F　F F　　Lewisian Gneiss

B

Figure 6.2: Drawing of (A) and section through (B) Suilven, Scotland. The section of B measures 731m from the west by northwest end to the east by southeast end; F= fault. Note the way a quartzite capping has preserved this astonishing dome of horizontally bedded sandstone above a rough rocky basement of eroded gneiss dotted with marshes and lochans (after photograph and Stephens, 1990, page 11).

and undulating surfaces. A notable accumulation occurs, for instance, on Fyfield Down (Wiltshire). They were sometimes called 'greywethers' because of their resemblance to sheep. They are the main rocks used in the construction of Avebury and Stonehenge (**Figure. 6.3**).

Iron influences the colour of sandstone. Where the iron is in the FeO (ferrous) state due to waterlogging and lack of oxygen its colour is blue. Where it is in the oxidized and hydrated state ($Fe_2O_3.2H_2O$) it is yellow and called limonite, and where in the oxidised but non-hydrated state (Fe_2O_3) it is red and called hematite.

Red colours in the landscape therefore indicate relatively dry conditions with abundant oxygen. This explains the general reddish coloration of deserts and has given its name both to the Old and New Red Sandstones of Britain. Yellowish brown colours indicate conditions where both oxygen and water are abundant, as in recent alluvial deposits. It is, for instance, noticeable in deserts that sand in river beds is yellow, but when it is blown out into dunes they become progressively redder with distance from the river source and the time since they were first blown from it. This seems to be due to the gradual conversion of the limonite coating on the sand grains into hematite. Bluish grey colours indicate waterlogged situations where bacteria have extracted oxygen from ferric oxide. Soils often show

Figure 6.3: Stonehenge's large uprights are sarsens.

a mottling of all three colours, indicating periodic alternations of aeration and waterlogging.

Sandstones can also be coloured black by free carbon, blue-grey by pyrites, and pink or purple by manganese. Calcareous sandstones are lime-cemented and usually whitish or greyish.

Valley heads in sandstone areas may be of three types: *spearhead* where a vertical fracture concentrates the water; *amphitheatre* where springs follow and enlarge the outcrop of a horizontal fracture, and *spongy* where water emergence is diffuse or from above each of a series of impermeable layers called *perched water tables*.

Permeability increases with the square of particle size, and high permeability means high erosion resistance so coarse sandstones tend to be more competent than fine ones.

Surface weathering is by granular disintegration followed by the peeling of thin weathered layers from the surface. Its rate increases with the vulnerability to weathering of the cementing matrix. The result is that sandstone landscapes in humid regions show domes and rounded slopes. By contrast, wind erosion in deserts can create angular forms. These include *pillars*, flat-lying *pancake rocks*, *mushroom rocks* where a wider rock is perched on a narrower stand, and castle-like forms which sometimes assume fantastic shapes. Isolated outliers which become rounded are called *gour* (singular gara) Small sharp-crested rocks formed by and aligned with, the prevailing winds are called *yardangs* (**Figure. 6.4**).

Sandstones are generally rigid. Slight bending through tectonic action or undersapping can cause rectilinear fracturing. Such fractures decrease inwards from the surface and tend to be more widely spaced in thicker rock layers. It is estimated that 80% of water movement within sandstones is along fractures. Weathering follows this, separating the rock into quasi-cubical blocks which increase in size inwards from the surface. In hot deserts, intense heating and cooling in the presence of moisture accelerates this process, and wind-deflation removes the finest fragments (**Figure. 6.5**).

Tafoni can develop on exposed faces of sandstones, usually through weathering behind a hardened crust. If circular and clustered these are called alveoles. Sometimes tafoni take the form of *honeycomb weathering* which, as its name suggests, is

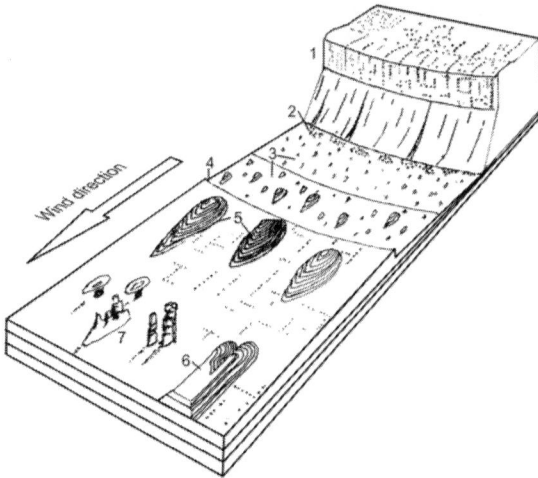

Figure 6.4: Wind-sculpted sandstone forms from Tibesti area: 1 escarpment; 2 cliff foot lake-bed type deposits. 3 small yardangs in depression; 4 rise in ground; 5 yardangs on higher level with alternating crests and corridors; 6 low plateau suffering erosion; 7 mushroom rocks, castle-like forms, and pinnacles. Note that forms 1, 2 & 4 more completely reflect their fluvial origin and that the effects of wind erosion increase with distance from the escarpment (after Mainguet, 1972, page 356).

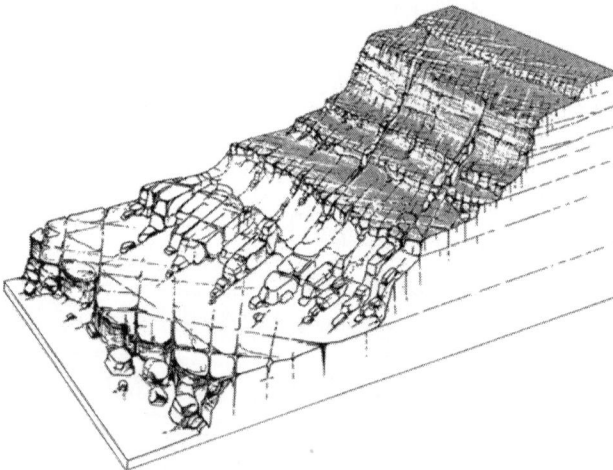

Figure 6.5: Schematic view of wind-eroded sandstone desert in the Sahara showing how fracture frequency decreases with depth and bed thickness. This reflects the fact that the greater the surface exposure the greater the cracking and deflation (source: Mainguet 1972, page 309).

a regular arrangement of fine pits a few centimetres apart in a pattern controlled by the bedding planes and jointing in the rock (**Figure. 6.6**). Continued weathering leads ultimately to the detachment of the perforated crust. When exposed to wave attack on a coast sandstone can form vertical cliffs and isolated pinnacles (**Figure. 6.7**)

Sandstone composed of fine sand particles, called *siltstone*, has a greater ability to maintain steep slopes. This is because silt particles, being smaller than sand, have a greater total surface area per unit weight and therefore a greater tendency to cohere. Unlike sand and gravel, they are not fragmental but have a flat platy form, causing them to interleave. This enables them to form stable vertical slopes with accompanying features such as alcoves, caves, and natural bridges.

The generally high permeability of sandstones makes for a "coarse" drainage density, with streams relatively widely spaced. This contrasts with the finer densities found in clay

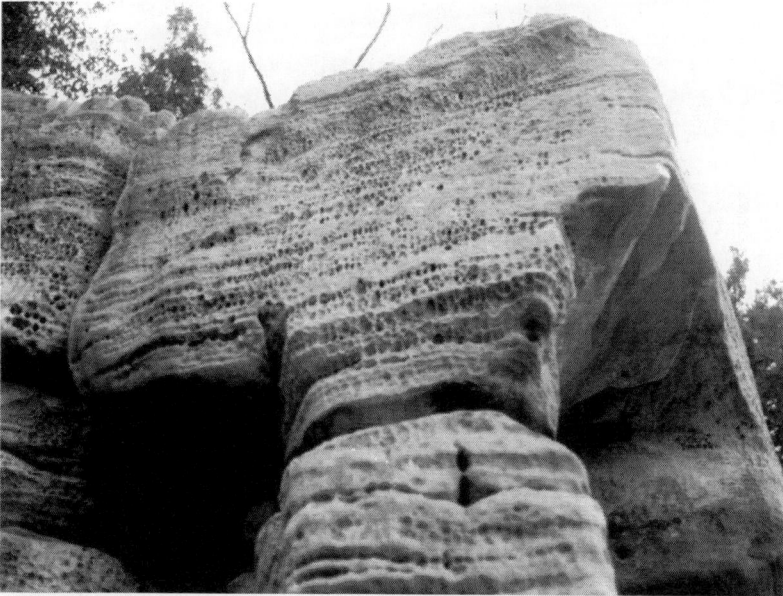

Figure 6.6: Honeycomb weathering in Wealden Sandstone near Crowborough, Sussex.This is particularly characteristic of sandstone. The cause is debated but it is suggested that a natural cement infilling internal joints is harder than the main rock mass, channelling erosion to give this mosaic of recessed hollows surrounded by projecting ridges.

Figure 6.7: Sandstone cliffs due to marine attack: the Old Man of Hoy (Orkney) on Old Red Sandstone (after Geikie 1887, page 292, Figure 67).

country and the disorganized part-subterranean drainage systems on limestones.

Siliceous materials yield a detritus of silt, sand, and gravel. Where there is little vegetation cover the wind can deposit the silt as loess (see chapter 17) and the sand as dunes. The less mobile gravel is left behind as a *lag deposit*.

Sandstones and their detritus give rise to fast-draining but generally rather infertile soils poorly suited to cultivation. In Britain they produce a distinctive scenery - common land, pinewood, and moorland. In the Italian Alps some hard sandstone beds flake off and can be used in building (**Figure. 6.8**).

Figure 6.8: Sandstone quarry, Pra del Torno near Torre Pellice, Italy. The rock slabs are usable for building with little extra cutting. The model men are illustrating the quarrying method.

CHAPTER 7

Calcareous Sediments

Materials and Processes

Calcareous sediments give some of the most extraordinary geographical features on earth. Limestone and chalk are composed of slightly soluble calcium carbonate. They always contain some magnesium carbonate which is about 7 times as soluble. Where this is an important component the rock is called dolomite. Gypsum (calcium sulphate) is also usually present which is about 145 times as soluble as calcium carbonate. These solubilities are much increased when organic acids are present. It is this solubility which gives a particular character to calcareous landscapes.

Limestones can be broadly subdivided into *sparites* and *micrites* depending on whether the grain size is larger or smaller than 0.01 mm respectively. In general, micrites are more porous and more jointed; so more easily weathered and dissolved. It is thought that this causes them to generate lower cliffs and crags and wider and shallower valleys. In the Ingleborough area of Yorkshire, for example, the Great Scar Limestone, a sparite, gives rise to the highest hills and steepest slopes. Where limestone has experienced intense pressures it loses all trace of its granular character and becomes *marble*.

Limestones are permeable and chalk even more so. Rain water concentrates in low places, and sinks in along bedding planes, joints, and fissures, dissolving the rock as it goes. The result is that there is little surface runoff and surface streams are far apart or absent altogether. Because the rock is carried away as a solute in water rather than as a suspension of particles, huge quantities of rock can disappear through very nar-

Figure 7.1: Limestone escarpment: Salève, France. Such a vertical scarp face is often characteristic of limestone as is the absence of detrital material at the cliff foot. This is mainly due to the rock's strong internal cohesion, and its tendency to lose material more from solution that from fluvial transport. Note the cave which is the tiny black square just below the trees on the summit.

Figure 7.2: Outer slopes of a dolomitic mountain above Pra del Torno, near Torre Pellice, Italy, whose steepness is partly due to the presence of significant amounts of magnesium carbonate with its high solubility. The light colours are due to the absence of dark ferric minerals.

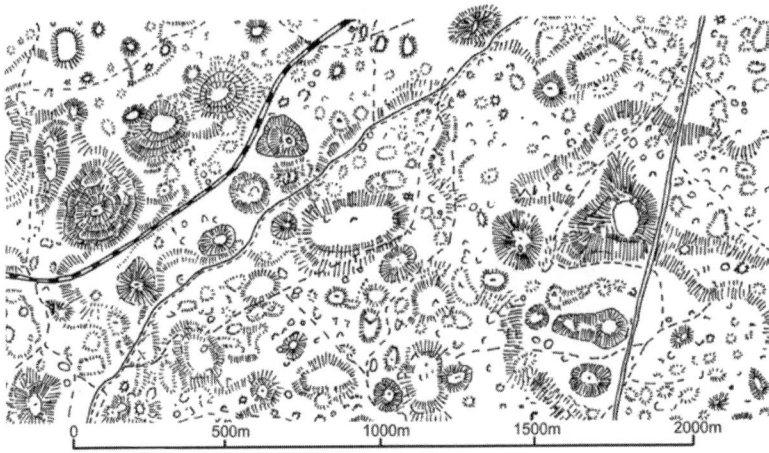

Figure 7.3: Plan projection of the doline landscape south of Sessana, former Yugoslavia, showing the wide variety of sizes and depths of the dolines (after J Cvijic in Sweeting 1981 page 28).

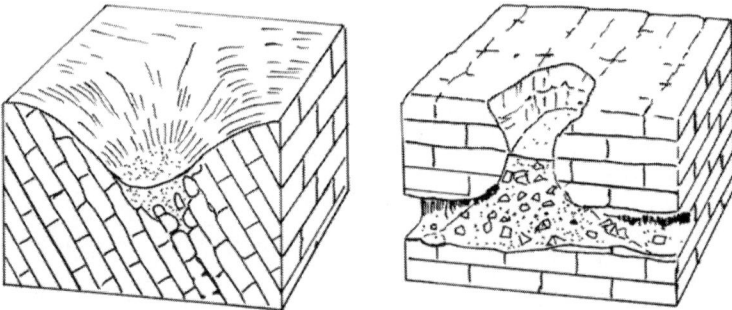

Figure 7.4: Two sorts of doline: left: solution doline, formed by water dissolution down bedding planes, and right: collapse doline due to the dissolution of a particularly vulnerable horizontal bed (after Bloom 1978 page 150).

row channels leading to often remarkable systems of caverns, ravines, boreholes and sinks. Similarly, in positions where lime-charged water evaporates it leaves solution deposits.

Mountain Areas

In mountain areas, limestone, especially where dolomitic, gives steep jagged landforms with wall-like precipices resembling

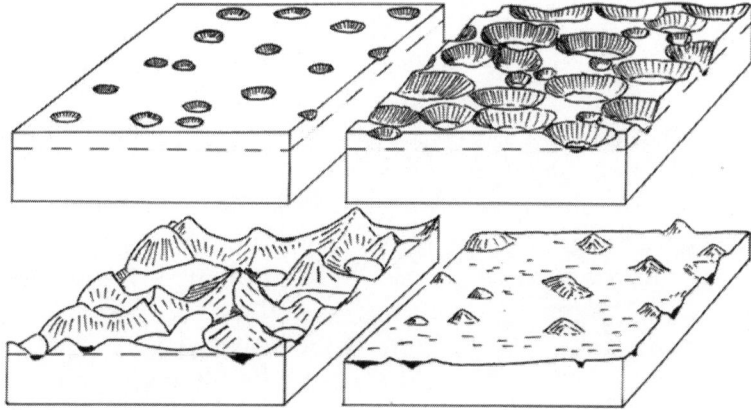

Figure 7.5: Model of the karst cycle showing stages in the aqueous dissolution of limestone deposits: top left, young karst; top right, adolescent karst - dolines larger and beginning to coalesce; bottom left, mature karst - 'cockpit' landscape; bottom right, old karst. Land has become even again between the cockpit hills (after Sweeting 1981 page 58).

ruined masonry (**Figure. 7.1**). Transverse and vertical markings indicate bedding planes and fissures respectively. The colours range from cream to grey. Climbing is difficult because of loose stones and insecure hand- and foot-holds. The Dolomite mountains of northern Italy are an outstanding example, closely similar to the mountain shown in Figure 7.2.

Karst

In humid climates, limestone terrain shows a network of surface depressions and eminences collectively known as *karst* after the type area in Dalmatia. They result from the concentration of surface water at low points. This gives dry minor valleys and patches of bare rock which may be honeycombed and fluted by rain water under a thin turf often providing good sheep pasture. The water dissolves the underlying rock, leading to collapse and further lowering. Surface depressions a few metres in diameter and several centimetres deep appear, called *kamenitzas*. These develop initially under standing water which deepens the hollows by solution. Then the sides, in contrast to the smooth floor, become increasingly fluted or fretted by free-flowing water. Basin-shaped, sediment-floored depressions more than 2-3 metres in diameter are called *dolines* (**Figure 7.3**). Two different types are recognized, according to origin. *Solution dolines* are due to solution of the underlying bedrock, *collapse dolines* to collapse of underlying caverns (**Figure 7.4**). Either can be bowl-shaped, funnel-shaped, or well-shaped. In general, the deeper the doline and the steeper its sides, the more important has collapse been in its formation. Where dolines are large as when due to the coalescence of two or more, they are *uvalas*. If the doline is large and elongated with the long axis parallel to the structural grain and floored with alluvium, it is called a *polje*. It is an *open polje* where it is drained by surface water, a *closed polje* if drained by swallow holes. An isolated residual hill on the floor of a polje is a *hum*. In arid areas solution hollows on limestone plains are much shallower but can be more than a kilometre in diameter. In North Africa they are called *dayas*.

Karst evolution goes through clear stages (**Figure 7.5**): - youth, adolescence, maturity, and old age. First, dolines and kamenitzas develop on an original surface. Then they increase in numbers and enlarge enough for some to coalesce. In the final stage the intervening areas become isolated and low standing above the residual plain which once was the floor of doline formation.

In tropical climates destructive solution of a limestone landscape can advance so far that the bulk of the material has disappeared. This leaves residual land forms known as *cone-karst* or the more mountainous form - *tower-karst*. Both consist of residual masses of limestone rising precipitously from a flat valley floor or plain They are usually due to faulting in the limestone which isolates it into blocks which are then undersapped by the water in which they stand. The 'stone forest' area near Giulin, China (**Figure 7.8**) and the pinnacles standing up in Halong Bay off the coast of Vietnam are well-known examples. Cone-karst is terrain with numerous cone-like hills, sometimes separated by deep, quasi star-shaped hollows called *cockpits*. These are due to a fluctuating water table acting on alluvium-floored dolines. The 'Cockpit Country' of Jamaica is the type area. On a smaller scale, *polygonal karst* is a network of residual calcareous mounds separating closed depressions.

Other types of karst include *glaciokarst*, produced by glacial scour followed by post-glacial solution, initially by meltwater

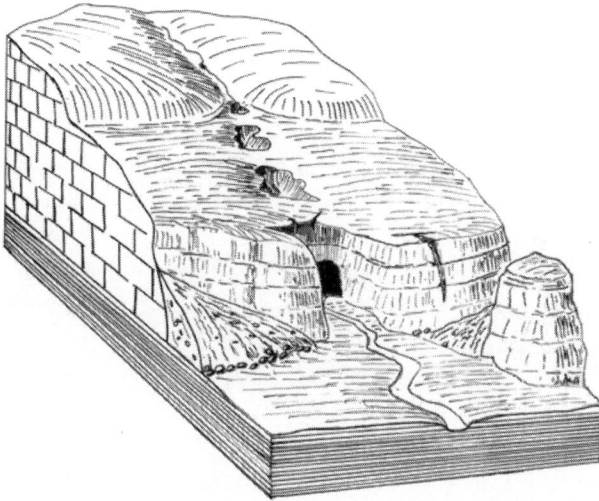

Figure 7.6: Block diagram illustrating swallow holes, limestone caves, underground watercourse, a gorge due to the collapse of a cave roof and natural bridges (after Shepherd 1952 page 81).

Figure 7.7: Stalactites hanging from the roof and stalagmites rising from the floor. The cause is that lime-charged water percolates into the cave from above and then the dripping water evaporates both before and after it falls to the floor. Aide memoir: stalactites 'hold tight.' (after Evans 1953 page 57).

streams. *Fluviokarst* consists of landforms due to the combined action of rivers and concentrations of surface water.

Where the limestone is cohesive and has been exposed to surface water for long periods, streams may go underground through apertures called *swallets* or *swallow-holes*. They then dissolve their way along natural joints into deeper courses in the rock until stopped by an impermeable bed or standing water. Then they move laterally, increasing solution and joining underground streams which may link networks of caverns. When such an underground river re-emerges at the surface, it is called a *lost river* after one with this name in New Hampshire, USA.

High narrow shafts in the roofs of the caverns are called *avens* whether or not they connect to the surface. If they do, the English term *pot-hole* is often used. The underground stream may emerge as a *rising*: a spring sometimes backed by a spring-head alcove. When the roofs of caverns collapse, the stream appears in a *gorge* or, when particularly narrow and deep, a *chasm*, over which *natural bridges* may survive (**Figure 7.6**). Narrow chasms due to solution along joint planes or fault lines are sometimes given the Serbo-Croat term *bogaz*.

Caves are common in limestone country both at the surface and underground. They occur along joints and bedding planes where water penetrates, or where it flows alongside. In the latter case, the cave walls may exhibit depressions. Where these are dish-shaped they are called *scallops*, where elongated, *flutes*.

Water dripping from the roof of caves evaporates and deposits its lime content as *tufa*. This forms *speleothems*, either as *dripstones* or *flowstones*. Dripstones can be subdivided into *stalactites*, *stalagmites*, and *helictites*. Stalactites hang from the roof and stalagmites build up from the floor of caves (**Figure. 7.7**). These can assume fantastic forms as in the Wookey Hole caves in England and the Carlsbad caverns in New Mexico. Helictites are an erratic and curved form of stalactite occurring where the drips do not fall clear but the water precipitates lime as it evaporates on its way down a sloping surface. Flowstones show the direction of the depositing water.

Limestone also has special patterns of erosion along coasts as a result of solution and bioerosion. The patterns resemble

Figure 7.8: Tower karst, called by the Chinese 'stone forest'. The dramatic landscape formed by tower karst is captured by a genre of Chinese watercolour painting of which this is an example.

Figure 7.9: Malham Cove limestone pavements, Yorkshire. The level areas are called clints, and the intervening clefts, grikes (after Whittow 1986 page 83).

those in other sedimentary rocks with vertical or undercut cliffs and *blow-holes* (Scottish *gloups*). The latter are vertical or near-vertical clefts in a coastal cliff linking a sea cave with the cliff-top through which it ejects columns of spray. They are formed by wave erosion along a fault or joint. Although sea water is generally saturated with $CaCO_3$ and so does not dissolve much limestone, wave-cut benches are vulnerable to boring by algae and sponges. When small pools develop, their attack is accelerated, often leaving a rough sharp surface.

Detail of Rock Surfaces

Individual calcareous rocks show a variety of dissolution hollows ranging in size from a few millimetres to more than a metre in diameter. They are known collectively by the German term karren, and also as lapies. On level surfaces these follow faults or fractures as *fault-karren* or *fracture-karren*. Pavements of bare rock separated by rain-enlarged cracks are known in the north of England respectively as *clints* and *grikes* **(Figure. 7.9)**. These may have a layer of loosened chippings

Figure 7.10: A chalk valley near Devil's Dyke, Chiltern Hills. Note the dry bottom and the gentler and more rounded slopes when compared with limestone. There is little soil development on such slopes so that they easily lose vegetation cover. The white marks show areas of bare chalk.

or flakes called *shillow*. The grikes may become enlarged into narrow run-off channels called *runnels* or develop into pits. Rounded hollows at their intersection are called *karren-rohren*. The clint surface may be weathered into a multiplicity of small sharp peaks, called spitskarren. On sloping surfaces runoff water can form parallel flutings called *scallop-karren* or *killen-karren* if the edges are sharp, or *rillen-karren* if they are not.

Sloping limestone exposures are subject to other forms of weathering. Where a hardened surface overlies a weakened underlayer tafoni are common, especially in coarse-grained rocks and where summers are hot and sea-borne salt present. Water breaches the surface carapace and then weathers laterally underneath it. This may go far enough to detach parts of the carapace, giving rise to new crusting underneath. Further weathering creates niches and caves.

In warm climates, dissolution is more active. Here, when calcareous boulders or stones containing concretions or veins

Figure 7.11: Chalk coast west of Birling Gap, Sussex. Note crumbly appearance and the two horizontal flint bands. (arrows).

of silica are subjected to dissolution, the less soluble silica is left behind and often stands proud of the surface in small ridges or peaks. On the smaller scale of single pebbles, a fine solution-pitting of the surface resembling worm-casts is known as *vermiculation*. In arid areas, limestone surfaces sometimes show a maze of fine cracks resembling breadcrust, thought to be due to the exploitation of the cracks by water.

Chalk

Chalk gives somewhat different landforms from limestone. Because it is softer and less cohesive, slopes tend to be gentler and the landscape more subdued (**Figure. 7.10**). Valleys are more open, almost always lack a surface river or stream, and are normally filled with detritus. In formerly glaciated areas these valleys are thought to have originated from sapping by ground water held up above a frozen layer. In the Chilterns of southern England, west- and south-facing slopes tend to be gentler than north- and east-facing slopes due to greater erosion as a result of greater exposure to weathering.

Escarpments in chalk can be steep, especially along the coast where undercut by waves (**Figure. 7.11**). Elsewhere they are often interrupted by shelves wherever there is a significantly harder bed. In England these beds are due to nodular chalk: the Totternhoe Stone in the Lower Chalk, the Melbourn Rock in the Middle Chalk, and the Chalk Rock at the base of the Upper Chalk. Their topographic effects are most clearly seen in Cambridgeshire and the northern Chilterns.

In humid climates chalk escarpments often have shallow vertical incisions called *frets* which may be regularly spaced. Undersapping may steepen the footslopes. Of the three divisions of the English Chalk: Lower, Middle, and Upper, the first has the highest proportion of insoluble materials and is underlain by the impermeable Gault Clay. Water which has percolated through the Chalk to emerge at this geological junction generates *valley-head springs*. These cause the escarpment to retreat by slumping, leaving deep *coombes*. There are examples in the Chalk escarpment overlooking the Vale of Aylesbury, and the *hangers* at Selborne and Hawkley, but the most extreme is probably the steep sided, basin-shaped Manger at Uffington in

Berkshire. Such hollows and surrounding footslopes are filled with *coombe deposits.*

Chalk landscapes in Britain north of the Thames have a covering of glacial drift dating from the Ice Age. Some areas have experienced *cryoturbation* (churning and heaving of the soil and subsoil caused by frost action). This locally gives geo-metrically patterned ground in which shallow chalk alternates over a distance of a few metres with upchurned drift. Locally it has some admixture of loess, also dating from the Ice Age. The unglaciated areas south of the Thames have a thin soil but with bare chalk in patches. The vegetation in both areas is normally turf. Where drift is present, the soil is thicker and stonier and valleys sometimes have surface water.

Secondary Lime Accumulations

Where lime-bearing water warms, evaporates or experiences a rise in pH, it precipitates limestone as *tufa,* usually some dis-tance from where it was picked up. This precipitation takes various forms. Sometimes it coats the landscape with a layer of calcrete which may take the form of. flowstone where there is a distinct flowage pattern, notably in the form of waterfall curtains. Where deposited by hot springs it is called *travertine.* Artesian groundwater sometimes forms lime-rich or gypsum-rich *spring-mounds* around the point of emergence. Where de-posited lime accumulates around the rim of a pool as on the floor of a cave in karst terrain, it is called *rimstone* and the pool a *rimstone pool.*

Soils

All calcareous rocks give a high pH to their overlying soils. In humid regions this almost always gives them a richer veg-etation and a more prosperous agriculture than is found on more acidic rocks in the same climate. A notable example is the relative fecundity of the White Peak of Derbyshire on lime-stone dating from the Carboniferous geological stage, called Carboniferous Limestone, when compared with the poorer vegetation and soils of the Dark Peak on the acidic Millstone Grit.

Uses

Limestone can form a good building material. Portland Stone was used in many important buildings including St Paul's Cathedral, Westminster Abbey, the British Museum and the National Gallery. Dolomite is harder and has an attractive white appearance.

CHAPTER 8

Shales and clays

Silicate and aluminosilicate minerals form the commonest surface sediments of all. They break down into clay particles. When transported by running water they remain suspended until reaching lakes or still water offshore, where they deposit as mud. When buried this compacts into mudstone and shale.

Because clay particles are finer than sands and silts, their sediments have a higher water-holding capacity but a lower permeability. This makes them slower to saturate but also slower to dry out. Shale landscapes in humid climates have much surface water in streams, ponds, and lakes and a more dense drainage network called a 'finer drainage texture'. They drain slowly and low ground is often waterlogged. A vegetation mat increases infiltration and reduces runoff, thus somewhat reproducing the effects of coarser materials. The result is that vegetated shales and clays have a coarser drainage texture than their bare equivalents. When more indurated (hardened by drying) and in leaf-like form called *slaty* they give an undulating hilly country (**Figure 8.1**).

Where surface water is abundant it attacks the weakly cemented particles, making shales and mudstones disintegrate and erode easily. These rocks characteristically have gentle slopes and frequent slump features. Thus they generally form the lower ground between uplands and the bays between coastal headlands. Where interbedded with coarser materials they form the more easily eroded layers (**Figure 8.2**). Steep slopes can occur where outcrops underlie more competent caprocks which save them from reclining to their natural angle of rest. Shales outcrop over considerable areas in the

Figure 8.1: Terrain on slaty coherent rocks seen from above Barr, Ayrshire. Note that the area is a gently undulating subdued landscape. This is because these rocks are relatively soft and easily eroded, differing from limestone in being insoluble and from sandstone in being both less permeable and more vulnerable to surface weathering. The highest peak in the distance is Merrick, the highest point in southern Scotland (Source: Geikie 1887 page 292).

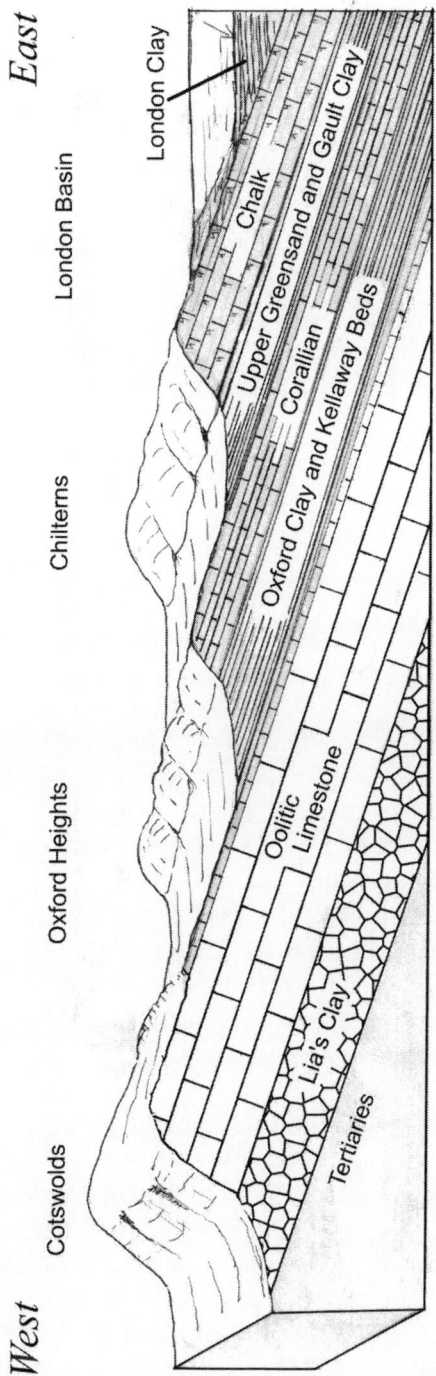

Figure 8.2: Schematic diagram of rock competence in southern England (not to scale). Note that the calcareous rocks, indicated by the 3 types of brick-like shading, vary in erosion resistance and the clays always form lower ground.

Figure 8.3: Badlands, SW USA. These are landscapes with deep dissection, ravines, gullies and sharp-edged ridges which have been created by fluvial erosion of relatively soft rocks in a semi-arid environment.

large geological basins around London and Paris and over the Flanders plain.

In arid climates with much bare ground, runoff is more intense. This cuts relatively steep slopes, gives a fine drainage texture, and quickly removes detritus. The extreme form is the intricately dissected broken ground called *badlands*, - a 'rough sea' of mounds and channels (**Figure 8.3**). The runoff concentrates in low places. When this evaporates it leaves a flat and level pavement of clays, often with a network of surface cracks and with some salt efflorescence. Such pavements are called takyr (**Figure 8.4**).

Even a small admixture of more competent material can make an appreciable difference in the landscape. A relatively narrow band of sand in a horizontal clay deposit can cause a significant shelf to appear when the area is eroded. An example is the shelf of the Middle Lias Ironstone below the Cotswold escarpment which consists of only slightly coarser particles than the surrounding Lias clays. The prominence of Mam Tor in the Derbyshire Millstone Grit is largely due to the intercalation of

Figure 8.4: Takyr in dry lake bed, SW USA.

bands of grit in the shales which form it. Another outstanding example is the Messines-Paschendaele Ridge which dominates the Flanders plain and has had high strategic importance in war. Its modest heights are due to an intercalated sand bed within the Flanders Clay (**Figure 8.5**). Enhanced competence may also be due to bands of stiffer, often calcareous, clays and marls. An example is the Chalk Marl of southern England which used to be called *clunch*.

Shale competence also increases where a river leaves an *alluvial train* of gravel and sand across a shale area. The materials in such trains are more erosion-resistant than the surrounding clays so that after prolonged erosion there can be an inversion of relief whereby the gravel and sand cap the high

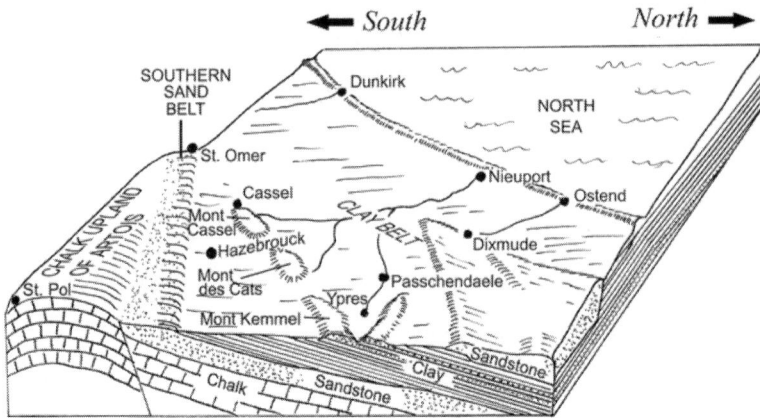

Figure 8.5: Schematic block diagram showing the importance of sandstone outcrops in clay landscape: the Flanders battlefield of World War 1. The Mont Cassel - Mont des Cats -Mont Kemmel ridge illustrates this. Mont Kemmel, though barely 150m high has been a key position in Flanders since Roman times, commanding wide views. In World War I six major battles were fought for its control.

ground while the clays are eroded to a lower level. Springs occur where these outcrop above the clays. Such springs often generate small streams.

Where chalk overlies less permeable sediments, a spring line occurs where downward-percolating rain water reaches the junction. This feature can be seen along the foot of the northern escarpment of the Chilterns which overlooks the Vale of White Horse. The villages are aligned along the spring line where the Glauconitic Marl expresses water from the overlying Lower Chalk.

Clay has many uses. It generally gives fertile soils. Where it contains calcareous material it is called *marl,* which is sometimes extracted to add to sandy soil to increase its fertility. Clay is also extracted for moulding and firing into bricks and ceramics.

CHAPTER 9

Duricrusts

In some landscapes the land surface is hardened into a *duricrust*. This results from impregnation with chemical compounds brought by ground water or surface flooding, followed by hardening on drying. Duricrusts protect old surfaces from erosion and are sometimes raised in the landscape as capping rocks over softer material. They are classified according to their hardening compound into *calcrete, silcrete, ferricrete,* and the rarer *alcrete, magnesicrete, dolocrete,* and *gypcrete.*

Calcrete, also called *caliche* or *croûte calcaire,* is material case-hardened with calcium carbonate (**Figure 9.1**). It is formed when lime is precipitated from percolating water. This can be caused by evaporation, an increase in temperature, a decrease in pressure, or a decrease in acidity. The lime mainly comes from calcareous rocks, although decayed vegetation, shell fragments, and windborne dust washed into the surface also contribute.

The typical site for calcrete formation is a valley bottom or lake bed. The impregnated material thickens towards the lowest points, so that if it is uplifted these can become the most competent parts, inverting the relief. Because it has the character of an impure limestone, it is subject to karst formation.

Calcrete is especially common in subtropical landscapes with an annual rainfall of between about 100 mm and 500 mm, such as North Africa and the Mediterranean. It can vary in thickness from less than a centimetre to as much as 60 metres. The climate of northern Europe is generally too wet for it although it is found in certain rocks in the geological record such as the Old Red (Devonian) and New Red (Triassic) sandstones in England.

Figure 9.1: A calcareous crust can form an overhanging cliff as here on a roadside cut in USA (after Soil Survey Staff p.210).

Magnesicrete and dolocrete, as the names suggest, are indurations caused respectively by magnesium carbonate and by mixed magnesium and calcium carbonates. They generally resemble calcrete and have the same semi-arid distribution. The main difference is that, because magnesium is less common and its carbonate more soluble than that of calcium, it is rarer and forms less conspicuous and less enduring landscape features.

Gypcrete results from impregnation with gypsum. It is softer and more soluble than calcrete and so is never prominent in the landscape and is insignificant in humid climates. It is commonest around and downwind of desert lakes, such as

those in Algeria and Tunisia. These lakes contain much gypsum and other soluble salts. When they dry out in the summer, they expose these salts in finely divided form. The wind
can blow them to places where they are trapped by vegetation
or moister conditions. Once trapped, surface water washes the
gypsum into the top few centimetres of soil where it hardens
into a pavement when it dries.

Silcrete is most characteristic of areas where high temperatures accompany alkaline soil waters because these allow silica
to be leached out of the soil and redeposited locally. This makes
it most characteristic of semi-arid areas because for different
reasons it does not much occur in the humid tropics or the cool
temperate zone. In the former, the soils are neutral to alkaline.
This breaks feldspars break down to silica and kaolinite, but removes them in drainage so that they are seldom in a position to
impregnate the surface. In the cool temperate zone, soil water
is generally acid and does not attack the silica, instead leaching away bases (metallic cations such as sodium, potassium,
calcium and magnesium) and sesquioxides (the oxides of iron
and aluminium). Silica does however appear here as concretions in calcareous materials and as a localized cementation of
porous rocks.

Even when it forms as little as 10% of a soil layer, silica
can cement it effectively. Silcrete can vary in thickness from
less than a centimetre to tens of metres, 50 metres having been
recorded in Zaire and Namibia. It is especially widespread in
Australia where the bedrock is largely siliceous. Where exposed to erosion, silcretes behave as indurated sandstones.

Silica concretions in limestone and chalk take the form
of *chert* and *flint* which, as we have seen, are common in the
British landscape (see **Figure 7.11**). Both are black, but chert is
duller in colour. They differ in that chert has flat, planar fractures while flint has conchoidal (shell-shaped) fractures. They
differ from the other types of duricrust in that that they occur
in separated concretions, not in a continuous sheet. Because
of their relative insolubility, fragments of both are sometimes
concentrated into a stony armour on the surface after the calcareous rocks containing them are reduced by dissolution.

Where ground water is charged with silica, evaporation
and an increase in the acidity of the soil water can precipitate

Figure 9.2: Flint cottage, Sussex. Flint occurs in chalk in bulbous lumps which its hardness and glass-like fracturing makes hard to shape into rectilinear blocks. It is therefore necessary to use brick around doors, windows and corners.

it in cracks and joints, forming quartz dikes and veins. These usually become harder than the rocks they inject, so that when weathered and eroded they stand proud of the surface as ridges or walls.

The process giving 'varnish' on stones in deserts appears to be a type of silcrete in the form of a thin opaline (glassy) crust. It is thought that the occasional rain penetrates into porous pebbles, notably sandstones. When it evaporates it draws out oxides of iron, manganese, and silicon, precipitating them at the surface of the stone. The first two give the dark brown coloration, the silica the outer varnish.

In humid tropical regions, iron and aluminium concentration mainly comes from the differential removal of surrounding materials under warm and generally non-acid conditions. Intense leaching of bases and silica, often to a depth of many metres, leaves high concentrations of iron and aluminium oxides at the surface. When exposed to air, the iron oxides harden irreversibly into a surface crust called *laterite* by a process which seems to include the formation of the mineral goethite. Where this crust is widespread and bare of soil and vegetation, it is called a *laterite shield*. Such shields cover considerable areas in the humid tropics, and in temperate regions may stand up as cappings in areas attacked by erosion.

In humid temperate climates the movement of iron in the soil is aided by *chelation*, a mechanism whereby organic molecules attach themselves to iron cations. The iron is re-precipitated as ferric oxide particles where there is lower acidity. These particles are then moved down the soil profile until held up by an impermeable layer, or until precipitated where better aeration in a sandy or gravelly layer evaporates the water. The resulting concentrations of iron oxides can harden into concretions, thin horizontal *iron pans*, or can act as a cement around sands and gravels. Their common occurrence in upland Britain is partly due to the removal of the ancient tree cover which led to soil acidification, mobilizing iron in the groundwater. The impermeability of these pans prevents through drainage and holds up water at the surface.

When ferricrete is exposed at the surface it resists erosion and can provide eminences in the landscape. Even a small iron impregnation can increase the competence of the rock which contains it, as can be seen by the way ironstone layers and iron-cemented conglomerates stand proud of the eroded faces of surrounding rocks.

Alcrete seems to be confined to the humid tropics. While ferruginous laterite develops on basic rocks such as greenstone, aluminous laterite develops on felspathic rocks low in iron. The absence of alcrete from humid temperate regions seems to be due to the relatively high solubility of aluminium oxides in the prevalent acid soils.

Uses:

Material from duricrusts is used for building mainly where it is the main hard stone available. Flint's separate particles are used widely in building although corners, windows, and door-ways require to be edged with stones or bricks which can be cut into rectilinear shapes (**Figure. 9.2**). Laterite is the basis of many buildings in tropical rain forest areas.

CHAPTER 10

Gravity-formed deposits

Loose materials which gravity has moved from their place of origin are called *colluvium*. They are of two types reflecting the nature of the space they leave behind: *weathering-limited* and *transport limited*. The first occurs where the parent rock yields weathered fragments which fall into an unsorted heap as soon as they are detached from the rock face. The second occurs where the weathered material cannot move downslope away from its source.

Weathering-limited materials falling from bare rock faces can be classified by size into slabs, rocks, stones, and grains. Where rock is coherent and its fractures widely separated, the detritus falls as slabs. At the other extreme, where the rock is soft and its greatest weakness is the bonding of individual grains, the detritus will be granular. Other forms of colluvium represent gradations between these two extremes. These fragments fall from a rock face as *scree*, also called *talus* (Figs. 10.1 & 10.2). Scree is sometimes differentiated from talus in that it can be on any slope while talus can only be below a bare rock face. Both lie against the hillside at an angle of rest characteristic of the materials concerned, with a longitudinal profile which is concave upwards, and with a tendency to be concentrated into distinct cones below reentrants in the source cliffs. The steepest angle for such materials to be stable is only 30-35°. Unlike alluvium, the largest fragments tend to fall furthest down the hillside. Where the talus is confined within a valley, it is called a *talus glacier* or *rock glacier* (Lobeck (1939) p 80). If the valley is steep it may be called a *rock stream*.

Transport-limited colluvium mantles the slopes which generate it. It is commoner in humid than in arid regions be-

Figure 10.1: Screes below an escarpment in the Atlas mountains , Morocco, showing the way they lie as detrital slopes below the rock outcrops from which their material is derived.

Figure 10.2: Wastwater screes from Middle Fell, Lake District. Wastwater Lake in foreground, Whin Rigg (535m) right rear. Note how erosion has generated the fragments forming the screes whose steepness is emphasized by the footslope's concealment in the lake. (after Wainwright, 1966, unpaged).

cause of generally higher rates of rock weathering. The slope profile depends on the scale and type of gravity-induced movements. These are generally initiated by water infiltrating into the detritus.

The downslope movements of detached materials are of two main types: *rock or soil creep* (**Figure. 10.3**) and *mass movement* (**Figure. 10.4**). Soil creep, often accompanied by the formation of rills, is the downward drift of soil under gravity. It mainly results from wetted surface layers sliding down over a subsoil where the plane of contact has been lubricated by water, often following the seasonal thawing of the surface layer. It is especially common under sub-Arctic conditions where the process is known as *solifluction*. Where the subsoil remains frozen and the soil slides over it, the term *gelifluction* (or *congelifluction*) is sometimes used. Solifluction reduces the thickness of loose surface material at the top of the slope and increases it towards the bottom. The pattern of downhill flowage is sometimes in the form of advancing lobes. The resulting hill-foot detritus is sometimes called *head*, or if it involves chalky material, *coombe rock*.

Mass movements are gravitational slides and slumps of unconsolidated materials, triggered by pressures along underground zones of weakness. They can be classified into three main types depending on the internal coherence of the material which in turn depends on their degree of water saturation. These are *slides, slumps,* and *flows*.

Slides mainly retain their internal structure. They follow underground surfaces and are lubricated by groundwater. The largest are called *landslides* and the smaller ones *landslips*. Where the underground surface is a concave curve, they are called *rotational slips*. All types, but especially rotational slips, may be multiple and may give rise to stepped *terracettes*.

Where the materials are sufficiently saturated with water to lose much of their internal structure they form *slump deposits*. These show some internal mixing and often have a broken hummocky surface. They may clog a valley or block a road. If they emerge on to a plain they may form a *detrital cone*.

Stone-banked terraces are a type of slump deposit occurring on moderate slopes, usually 10-25°. They are steps of fine soliflucted material flanked by risers of coarser material. They can

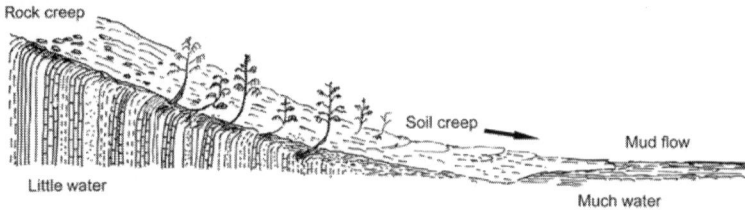

Figure 10.3: Rock creep, soil creep, and mudflow, showing how detrital materials become finer with distance from their source (after Lobeck, 1939, page 92).

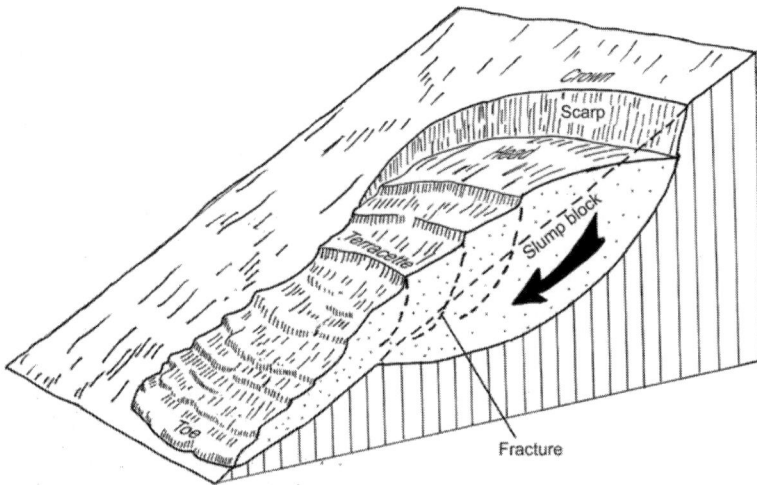

Figure 10.4: Mass movement: a slump flow. Note the crown, scarp, head, fractures, terracettes and toe (after Bloom page 178).

be up to 30 metres wide and 3-5 metres high. The stones appear to be loosened by frost and then heaved to the surface and migrate downwards faster than the fine material until arrested by slight changes in slope, vegetation, or rock exposure. They form a nucleus against which the finer material builds up.

Where the material is saturated with water beyond its "liquid limit" it behaves as a viscous liquid rather than a solid. It is called a *mudflow*, or if rapid, a *mudspate*. These occur suddenly when triggered. The pre-conditions are that a perched mass of permeable and non-cohesive material consisting of a mixture of different particle sizes from boulders to clay is suddenly saturated and collapses following an event such as a rainstorm or an earthquake. They are commonest in semi-arid areas subject to sudden and violent storms, but rarer where there is a mat of vegetation. They can be highly destructive. Where the mudflow emerges on to a plain its viscosity gives it a creeping movement terminating in lobes. The materials become churned so that the resulting deposit is thoroughly mixed and unstratified.

CHAPTER 11

Ice and glacier erosion

Erosion by ice and snow leaves indelible marks on the landscape. In temperate climates it is most often seen in mountain areas with a lot of bare rock. Surfaces are mainly weathered by freezing and thawing when wet which cause an alternation of expansion and contraction. When compacted, moving masses of snow and ice sculpt the surface by scoring, scouring, grinding, and plucking.

In higher mountain areas snow fills gullies and hollows. The gullies terminate abruptly at the upper end and are called *couloirs*. Snow lodging in hollows starts a process of both weathering and scouring called *nivation*. Runoff rain water or meltwater accelerates weathering under the ice which at these low temperatures has a greater ability to dissolve oxygen and carbon dioxide. The result of this weathering is the formation of shallow basins called *niches, nivation hollows,* or *cirques.*

Niches are small hollows or gullies, without exit lips, which can support niche glaciers.

Nivation hollows are larger depressions with a semi-circular plan but lacking a rocky threshold and the general armchair form of the cirque. The space between the bedrock backwall and the upper end of the snow mass is called the *randkluft*. Where the underlying material is limestone, the increased solution leads to *nivation karst*. This results in the formation of a backslope in the form of an arcuate scar which gradually recedes, thus widening the floor and increasing the snow-holding capacity of the hollow.

Cirques (Scottish *corrie*; Welsh *cwm*) are larger than nivation hollows and are gouged out by the ice and also differ from

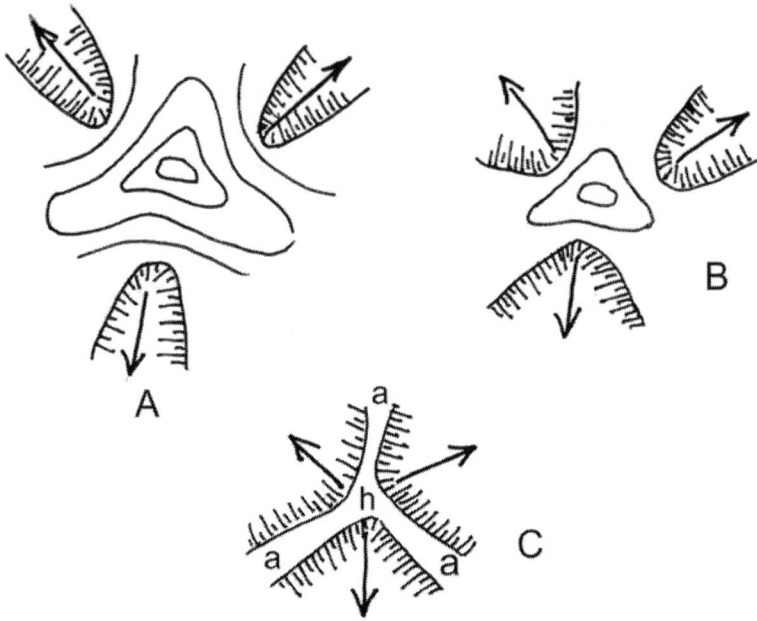

Figure 11.1: Schematic diagram in plan view showing stages A to C in the formation of arêtes (a) and horn (h) by backward development of cirques. The arrows indicate the direction of flow in eroding valleys. First, A: the cirques experience headward erosion, then B: they approach the summit, and finally: C: they leave a steep pointed horn with sharp arêtes leading up to it.

them in having a rising rocky threshold which gives them a general armchair form. When two cirques on a mountain erode backwards so as to intersect they leave a sharp separating ridge crest called an *arête*. Sharp steep peaks of Matterhorn type are formed when more than two cirques retreat so as to intersect **(Figures 11.1 & 11.2)**.

The snow in a cirque is compacted and recrystallized by succeeding snowfalls into material called *firn* or *névé*. When the cirque is full, the snow breaks away from the back wall leaving a separation crack known as a *bergschrund* **(Figures 11.2 & 11.3)**. The compacted snow and ice flows out of the cirque into the valley. over a terminal rock step. This steepens the gradient, accelerating the rate of ice flow and producing deep *crevasses* forming an *ice fall*. It then continues as a *glacier*.

Residual peaks left by backwearing cirques

Figure 11.2: High mountains illustrating horns which have come from the headward erosion of cirques: front to rear: Schalihorn, Zinal Rothorn, Matterhorn, Switzerland (from a photograph).

The glacier scours its floor and walls by two main processes. *Abrasion* is most active above and before obstacles where its flow is compressed. *Plucking* of rock fragments is most active below the obstacle. This is because the flow opens out and is aided by the re-freezing of water which had melted under compression above the obstacle.

Rocks held in the ice scour *striae* or *Sichelwannen* in underlying rocks. These are among the most reliable evidences of past glaciation. Striae are long parallel scratches oriented in the direction of ice movement. Sichelwannen are crescentic cavities with horns pointing in the direction of ice flow, resembling

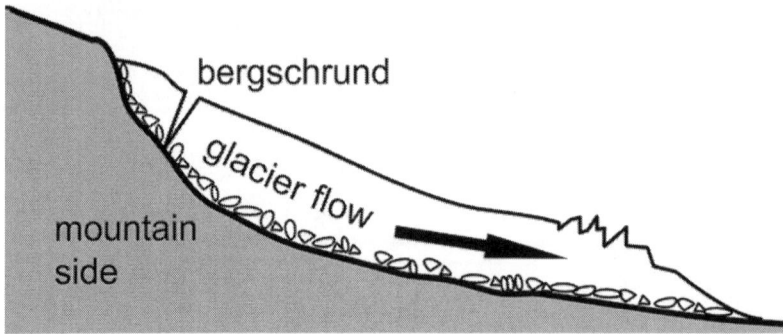

Figure 11.3: Detail of a bergschrund leading to a downward slide of ice on a mountainside.

flute or ripple marks fashioned by water on less resistant beds. They are thought to be due to differential corrasion in subglacial channels.

The lower end of the glacier, heavily charged with rock debris, is called the *terminus* or *snout*. Its location is determined by the dominance of melting over recharge. The point down the glacier where the two are equal is known as the *firn line*.

Glaciated valleys

Glaciated valleys develop a characteristic U-shaped cross-section with steep walls, a flattish base, and a straighter course than is normal in fluvial valleys. Lateral spurs are truncated and "battered". The floors of tributary glaciers are at higher levels than the main glaciers, so as to become *hanging valleys* after the disappearance of the ice (Figures 11.4 & 11.5). Glacial valleys may be scoured to great depth, but usually contain shallower rock bars, called *riegels*, downstream of the deeper stretches.

Streams of meltwater may cut V-shaped notches in the floor of U-shaped valley. These then form channels which may become sinuous and create interlocking spurs, introducing a fluvial component to the glacial landscape.

When valleys are drowned by the sea they become *fjords* (**Figure 11.6**). These may be very deep, Scotland's Loch Morar, for example, extending to 315 metres below sea level. Fjords have shallow thresholds and are usually deepest some distance

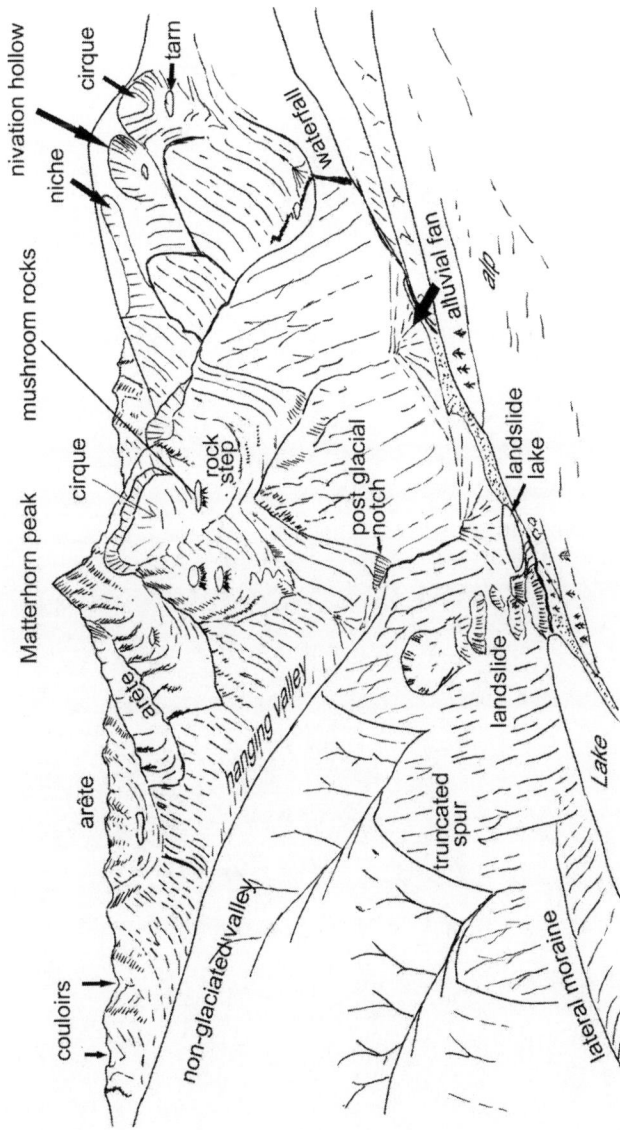

Figure 11.4: A mountain area after glaciation. Note the peaks, arêtes, cirques, niches, couloirs, U-shaped valleys, steps, moraines, and smaller features carved by the glaciers, and the landforms due to fluvial action after the retreat of the glaciers: tarns, landslides, notches, waterfalls, and alluvial fans. (after W M Davis 1906 & Lobeck 1939 page 262).

Labels in figure: nivation hollow, cirque, tarn, niche, waterfall, Matterhorn peak, mushroom rocks, cirque, rock step, alluvial fan, alp, post glacial notch, landslide lake, arête, arête, hanging valley, landslide, truncated spur, Lake, couloirs, non-glaciated valley, lateral moraine

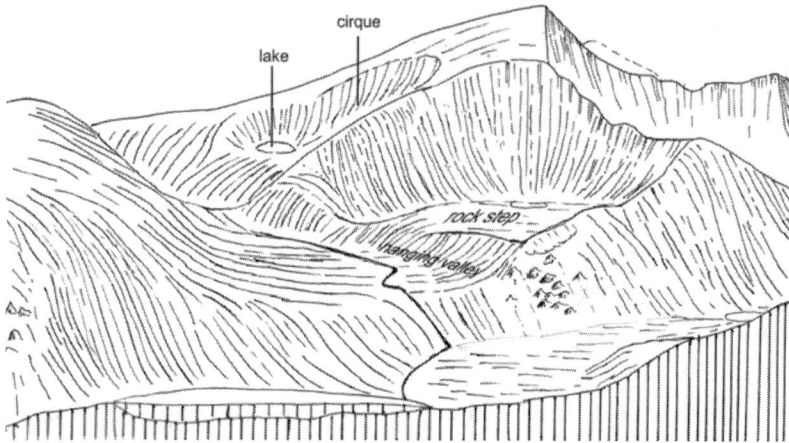

Figure 11.5: Mt Snowdon showing cirques, hanging valley, rock steps and lakes (after W M Davis 1906 & Lobeck page 264).

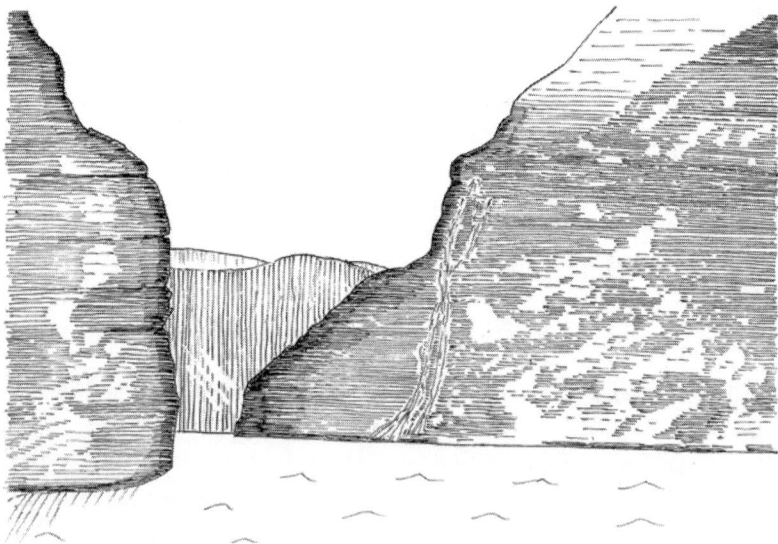

Figure 11.6: Geiranger Fjord and the Seven Sisters Falls, Norway. This is a former river valley drowned by the sea in a mountainous area showing very steep coastal slopes (from a photograph).

Figure 11.7: Section through the city of Edinburgh, showing the 'crag-and-tail' leading from the castle down the High Street to the Holyrood parliament building. The castle stands on a volcanic neck which was at times a nunatak during glaciation.

above their mouths. These shallower stretches give bars or islands across the fjords. These may be due to moraines (rocks deposited by the glacier) at the outer end of the former glacier, but more usually to subsequent deposition.

Ice-sculpted plains

Ice covering a plain forms an *ice sheet*, called an *ice field* if thick. Isolated crags rising above either are called *nunataks*. Where a nunatak protects an elongated ridge of weaker rocks, the removal of the ice leaves a landform called *crag-and-tail*, with the tail formed of glacial detritus. An outstanding example underlies Edinburgh Castle and the Royal Mile (**Figure 11.7**). Where the situation is repetitive it is called *a crag-and-tail landscape*.

Where a protruding rock is overridden and rounded by ice, it often becomes asymmetrical with an ice-plucked steeper side facing the oncoming ice and a gentler smoother stoss side as the ice passes over. Such rocks, when exposed after the retreat of the ice, are called *roches moutonnées* from their resemblance to sheep (**Figure 11.8**).

In level hard-rock country, ice-scour gives low rocky knolls (*knobs*) and ice-gouged depressions often filled with water (*lochans*). The combination of these is called *mamillated topography*. It is widespread in the granitic areas of northwestern Scotland and the Outer Hebrides, where it is called *knob-and-lochan country* (see **Figure 5.7**). It contrasts with somewhat similar terrain on glacial drift called *knob-and-kettle country*.

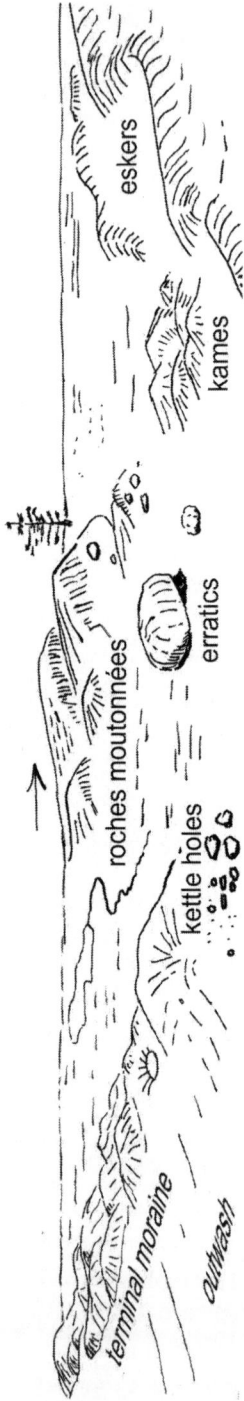

Figure 11.8: Types of glacial deposits left by continental glaciers. The terminal moraine is a detrital dump marking the limit of a former ice advance. The outwash behind it is due to deposition from the meltwater as the ice retreated. The kettle holes are hollows left when a block of ice encased in sediment melted. Roches moutonnées are rocks sculpted by overriding ice. They have asymmetrical form with the gentler abraded slope facing upglacier and the steeper plucked slope facing downglacier. Erratics are blocks dumped by retreating ice. Kames are steep-sided ridges or hills deposited by ice meltwater. Eskers are narrow sinuous ridges of sand and gravel deposited from glacial meltwater whose sinuosity is thought sometimes to be due to their having formed in channels beneath a melting ice mass (after Lobeck, 1939, page 298).

CHAPTER 12

Glacial deposits

Drift

Glacial drift includes all material deposited by ice or melt-water. It covers about 8% of the earth's land surface but nearly a quarter of North America and a third of Europe.

Ice-deposited material is known as *till*. It is generally an unsorted mixture of fragments of all sizes from clay to boulders reflecting the geology of its source area. It tends to show some orientation of stones with long axes in the direction of ice movement. It can be classified according to its mode of carriage and deposition by the ice. *Sublimation till* originates as a burden on top of the ice. If this has been reworked after first deposition it is known as *deformation till*. *Lodgment till* was laid down under the moving ice and *push-till* was bulldozed forward by the glacier. *Melt-out* till is the product of top melting of a debris-rich buried ice mass which retains the structures it had in the ice mass. Till which is compacted and lithified (solidified into stone) is called *tillite*.

Glacier deposits

Till laid down by glaciers has somewhat different landforms from that laid down by immobile ice sheets. As glaciers move down mountain valleys they scour materials from their beds and walls. These materials fall as a rain of detritus on to their margins, forming a *lateral moraine*. Where two glaciers meet, the neighbouring lateral moraines join to form a *central* or *medial moraine*. Morainic material within the ice is called *englacial* or if on the glacial floor covered by ice, *basal* (**Figure 12.1**).

117

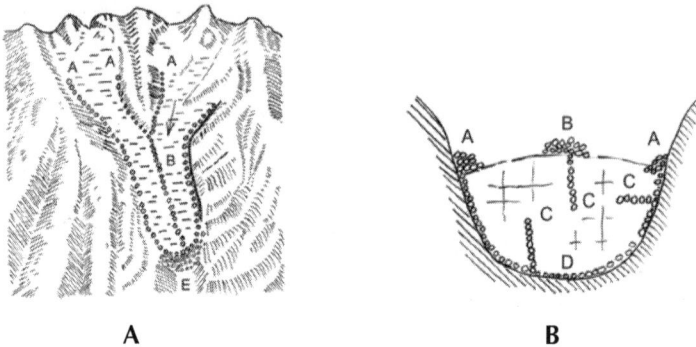

Figure 12.1: Position of moraines in relation to the glacier (a) and in cross section (b): A lateral; B medial; C englacial; D basal; E terminal, (after Gorshkov & Yakushova page 213).

Valley floor till, combining both *ablation till* and *lodgment till*, usually forms an undulating surface of low relief called a *ground moraine*. *Terminal moraines* mark the outward limit of glacial advance (**Figures 12.2 & 12.3**). They tend to curve downslope across the valley floor from its sides. When the ice retreats in stages it may leave a series of *recessional moraines* (**Figure 12.3**). These may appear as irregular crests and rows of piles of debris rather than continuous ridges. Even if erosion subsequently effaces the whole valley segment, we can sometimes infer the original position of the moraines from the alignment of their remnants on the valley sides.

Push-till forms *push moraines*. These can be distinguished from terminal moraines by having convex cross profiles with steeper slopes especially on the distal downslope side. A cut section will show internal thrusting and faulting due to the churning action of the ice which formed the moraine. Where the bulldozing action is due to the movement of surface ice into a lake during its winter freeze, the result is a *lake rampart*.

A glacier carries stones which can scrape striations in the underlying rock (**Figure 12.4**). Sometimes it scours parallel ridges and furrows in loose material called *fluted moraines*. The ridges can be 80-150 m long, a few centimetres to 2 metres wide and up to 30 cm high. By contrast, systems of small parallel ridges transverse to the direction of flow are caused by the episodic melting of glaciers at their point of debouchment into the sea or a lake. These are *cross-valley moraines* or *washboard moraines*.

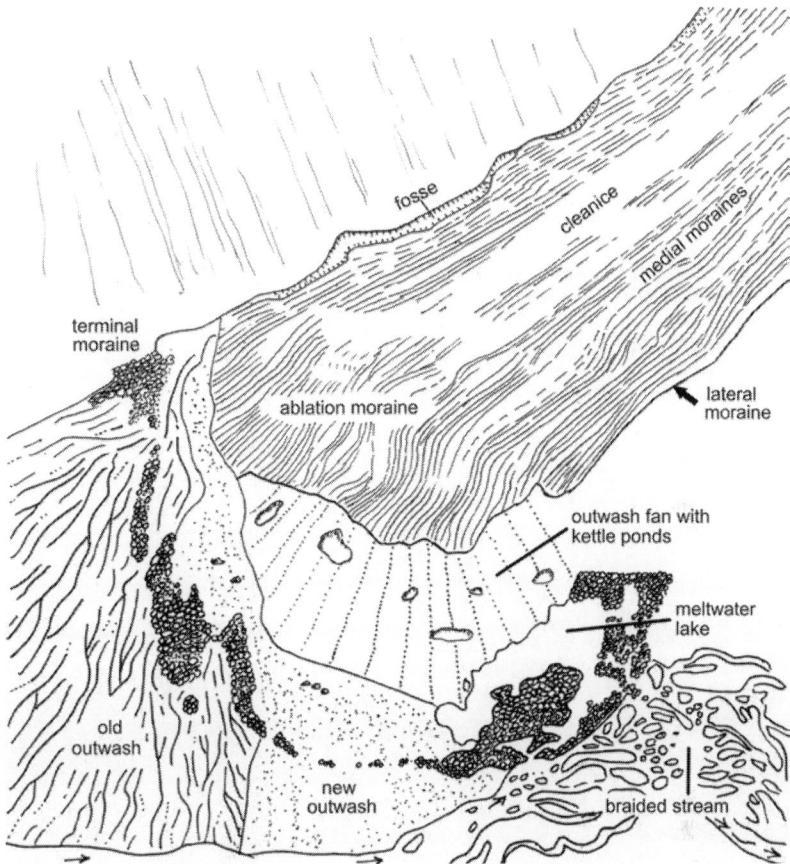

Figure 12.2: Terminus of a valley glacier in a mountain area. Note moraines: lateral when they are on the glacier's edge, medial when two laterals meet, ablation when morainic material is concentrated by ablation or melting. The fosse is the depression between the glacier and rock wall. The distinction between old and new outwash reflects the changing direction and pattern of meltwater flow. Kettle ponds fill kettle holes. (after Thornbury page 377).

Ice sheet deposits

The surfaces of stagnant ice sheets can show features so closely resembling those on limestone that they have been called glacial karst. They include *glacial poljes, uvalas, hums,* funnel-shaped *sink-holes, ice caves,* and *ice tunnels,* all of which can attain dimensions of more than a kilometre though are

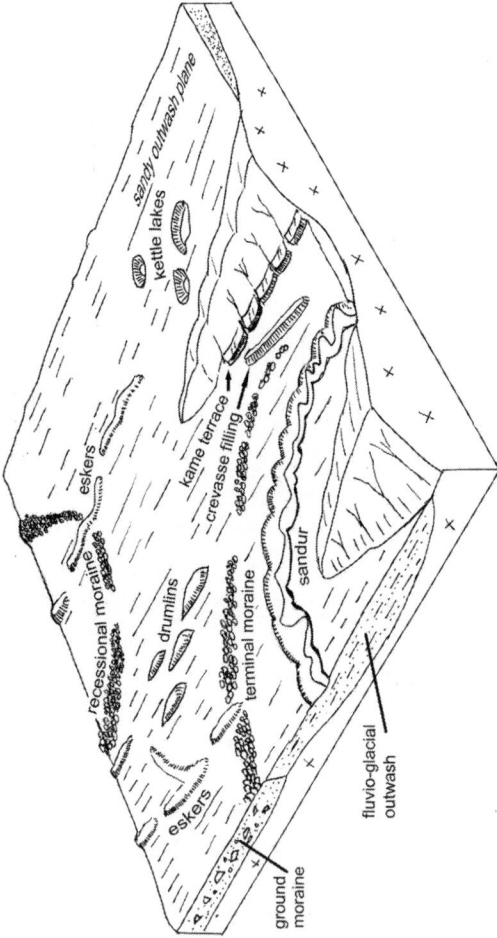

Figure 12.3: Idealized diagram to show the form of common glacial deposits. Moraines are deposited from ice, called 'ground' when deposited as they progress, 'terminal' when they mark the limit of glacial advance, 'recessional' when dumped in a pause in its retreat. Outwash is deposited from meltwater, forming kames in crevasses in the ice, eskers when the depositing channel is sinuous. The kame terrace is a flat-topped ridge or terrace of glaciofluvial materials between a valley glacier and the valley slopes. Crevasse fillings are materials dumped in crevasses in the ice which survive its melting. Drumlins are streamlined elongated hummocks of glacial material with the steeper slopes at the 'upstream' end. They may be due to the ice remoulding underlying loose material or depositing it around a rock or frozen drift. Sandurs are outwash plains of glaciofluvial materials carried out from the front of ice-sheets by meltstreams (after Thornbury page 396).

usually less. *Glacier moulins* are vertical walled cylindrical holes in the ice, sometimes due to accelerated melting under stones which transmit more heat than is experienced by the surrounding ice. Physical pressure causes ice to melt. It refreezes when the pressure is removed. The process is called regelation and is the effect that lubricates ice skates. Regelation aids a glacier's downslope movement as the ice melts around obstructions and refreezes behind them and forms a layer of water lubricating the flow. The regelation layer is the layer at the bottom of the glacier that is subject to this alternate melting and refreezing.

When ice melts relatively rapidly over a level surface, it leaves a *till plain*. This usually has an undulating or hummocky surface with occasional boulders, piles of stones, and depressions. Boulders brought far from their place of origin are called *erratics*. In Britain, some have originated as far away as Norway. Piles of stones are called *perforation deposits* when they have been dumped through melted perforations in the ice sheet. Depressions in the till formed from the melting of patches or lenses of underlying ice that is no longer being moved down by the glacier (called *dead ice*) are called *kettle holes*. These may fill with water and form ponds. *Knob-and-kettle* topography occurs where the moraine appears as a belt of knobby hills interspersed with basin-like hollows. Terrain of this type occurs widely on Cape Cod (Massachusetts). Terminal moraines tend to form arcuate patterns indicating that the ice advanced in a series of lobes. Where two lobes unite by curving back at right angles to the ice-front margin, the resulting form is an *interlobate moraine* (**Figure 12.1**) which when extended becomes a medial moraine (**Figure 12.2**).

Drumlins are hills formed of till and shaped like a half egg cut lengthwise with the stoss end facing up-glacier (**Figures 12.3 & 12.5**). They always lie in the zone beyond the terminal or recessional moraines, and can be up to 60 m high. The ratio of length to breadth ranges from 2.5:1 to 4:1, with the degree of elongation increasing with the rate of the ice flow which formed them. They often occur in groups or even swarms as in parts of Northern Ireland, northern England (**Figure 12.7**) and southern Scotland. Drumlins with a rock core are called *rocdrumlins* (**Figures 12.5 & 12.6**).

Figure 12.4: Glacial till resting on a striated surface (after Lobeck page 301).

Glaciofluvial deposits

Glaciofluvial landforms are laid down by meltwater from ice sheets or glaciers. The commonest are *kames, eskers,* and *sandurs.* Kames are steep-sided ridges or conical hills originating from crevasse-filling in the ice or from the dumping of deltaic material washed into lakes on top of ice which has subsequently melted (**Figure 12.8**). The materials are generally well sorted but poor-

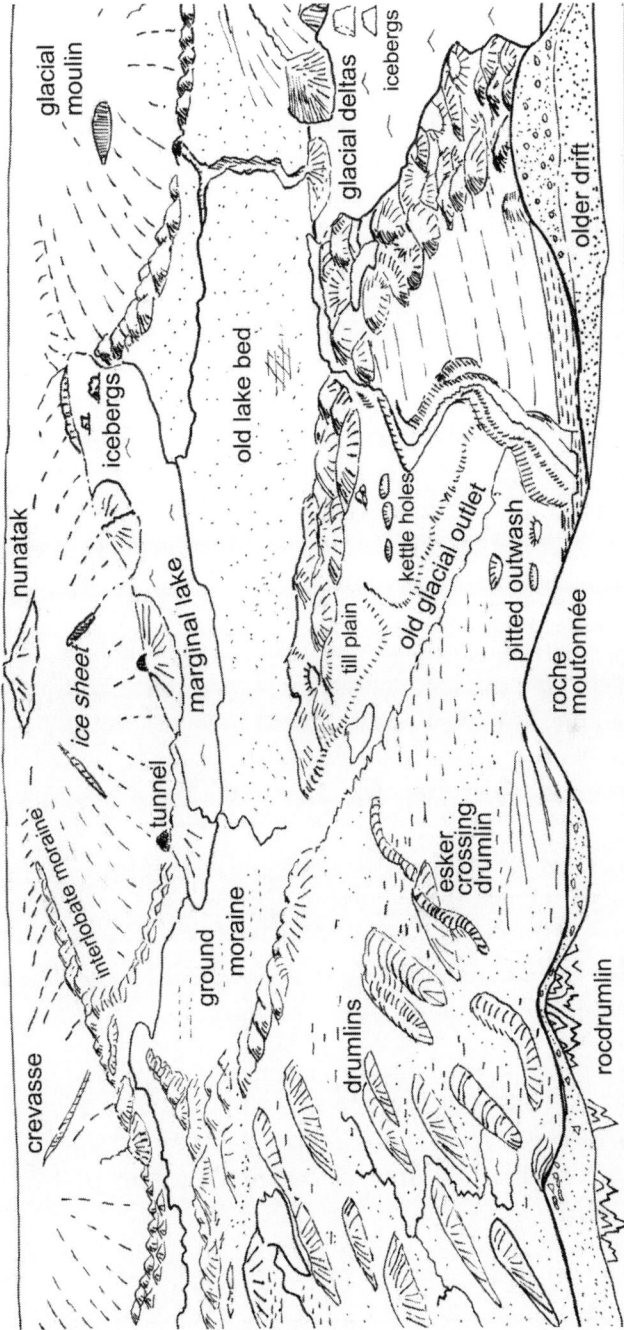

Figure 12.5: More types of deposit left by continental glaciers. Nunataks are isolated rocky peaks protruding above an ice sheet which have never been glaciated. The glacial moulin is a shaft or circular hole in a glacier down which meltwater pours. Rocdrumlins are rock features moulded by ice into drumlin shape. Till is detritus directly deposited by glacial ice and when this is level it is a till plain (after Lobeck page 298).

ly stratified, though *varves* (alternating bands of sand and clay usually due to varying seasonal deposition) are sometimes found. A group of kames constitutes a *kame moraine*.

A flat-topped ridge laid down by a meltwater stream flowing alongside a glacier in a valley is called a *kame terrace* (**Figure. 12.3**). This may contain kettle holes left by the melting of residual ice blocks isolated by the retreat of the surrounding glacier. An undulating landscape composed of kames or kame terraces and pitted with kettle holes is called *kame-and-kettle moraine*.

Glacial deltas (also called deltaic moraines) are deltas formed by a decaying ice margin at the edge of a lake (see **Figure 12.5**). They have a steep ice-contact slope at their upper margins, a surface gently inclined lakewards, and a lobate termination in the lake. They often contain kettle holes. Because of fluctuations in level, the glacial delta will have been formed by periodic alternations of silting and scouring. A characteristic sequence of beds is *bottomset-foreset-topset* (**Figure 12.9**). Where the delta comes at the termination of an esker it is called an *esker delta*.

Meandering meltwater deposits narrow sinuous ridges of partly stratified coarse sand called *eskers* or *osars* (**Figure 12.2**). It is uncertain whether they form under the ice sheet or at its margin as it retreats. An example can be seen near Holt, Norfolk.

Sandurs are outwash plains of glaciofluvial material composed of horizontally bedded gravel, sand, and silt and crossed by braided streams (**Figure 12.3**). Changes in the ice front may create moraines or ice-contact slopes which appear as ridges or steps across its surface. Where a sandur is pitted with kettle holes, it is called a *kettled sandur*; where it terminates in a lake, a *sandur delta*. Where the outwash is confined to a narrow line down a valley it is called a *valley train*.

Some water bodies at the edge of melting glaciers have received type-names. A *ribbon lake* is a temporary narrow body of water impounded between the front of an ice sheet and its recessional moraines. Meltwater streams debouching into such a lake

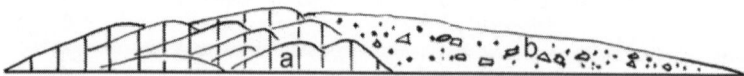

Figure 12.6: Drumlin cross-section: a) rock; b) moraine.

Figure 12.7: Drumlin field like a shoal of stranded whales, Hortonin-Ribblesdale (after Muir page 21).

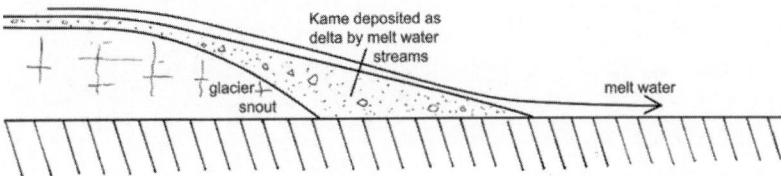

Figure 12.8: Kame formation at the glacier snout (after Collard page 215).

will form kame terraces. As ribbon lakes increase in size they coalesce to form *proglacial lakes*. A chain of lakes impounded by moraines in a glacial valley has been called *paternoster lakes*.

Periglacial features

Periglacial features are formed in environments close to permanent ice sheets or glaciers, though not actually moulded by them. They usually have a tundra climate and vegetation. The landforms can be classified into features due to ground ice, floating ice, ground patterning, nivation, solifluction, and gelifluction. Nivation is hill slope erosion by frost action, mass wasting and erosion by meltwater. Solifluction is the slow downslope flowage of water-saturated surface waste. Gelifluction is a type of solifluction associated with temporary or permanently frozen ground. Most periglacial environments are underlain by a permanently frozen *permafrost* layer and experience an annual freezing and thawing of the surface soil above this causing cryoturbation. They alternate between conditions of severely impeded drainage during the short summer and hard freezing during the long winter. The freezing is from above and compresses the wetted layer against the underlying permafrost. It is accompanied by the formation of downward-pointing ice wedges which churn the materials as they freeze. When the ice wedges subsequently melt fragments of surface material fall into the vacated spaces. This accounts for the upheaved and convoluted soil profiles often seen in periglacial areas.

Pingoes are dome-shaped hills built around cores of ice.

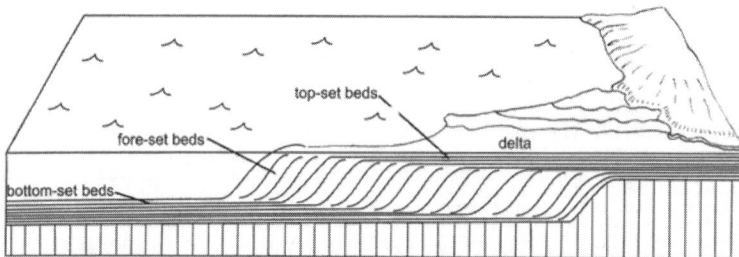

Figure 12.9: Formation of a delta showing top-set, fore-set and bottom-set beds (after Whittow 1984 page 204).

They can be 60-600m high in Alaska, Greenland, Siberia, and Canada. Where the eminence is only a few metres in diameter, - a blister in the soil above an ice lens, it is called a *hydrolaccolith*, and where covered with peat rather than drift and having ice in lenses rather than in a single block, a *palsa*. In all such ice-cored forms, melting dumps the materials and leaves a ring of detritus around a hollow centre. This combination of hollows and ramparts is sometimes jokingly referred to as an *ognip*, and can be a home for unusual flora.

Erosion at the edge of snow drifts forms *nivation hollows*. When materials fall on to the upper part of the snow bank they slide to its lower edge and form a bank of frost-shattered debris there. This is called a *protalus rampart*. It resembles a moraine except that, unlike it, it tends in plan to be convex towards the upper side.

Patterned ground

Patterned ground is a common phenomenon in unconsolidated surface materials in periglacial areas. The commonest types are *ice-wedge polygons, stone pavements, stone polygons, stone garlands*, and *stone stripes*. There is some debate about the origin of these features, but they are due generally to cryoturbation.

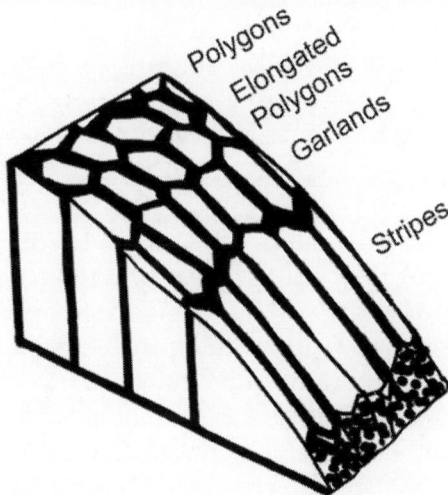

Figure 12.10: Idealized relationship between the angle of slope and stone polygons, stone garlands and stone stripes.

Ice-wedges can extend to 10 metres below ground and can be over 1 metre wide at the surface. They separate polygons which range in scale from about a metre to over 100 metres across and can be clearly seen on aerial photographs. It is thought that contraction under extreme cold causes the cracks. Snow may enter them and hoar frost form in them to such an extent that the thermal expansion due to the summer thaw is inadequate completely to close them. This leaves ice wedges pointing downwards which tend to increase year by year. An amelioration of the climate can thaw the permafrost, allowing surface sediments to slump into the voids left by the melting ice wedges, so preserving their forms as ice-wedge casts.

Related to ice-wedge polygons are forms which, seen from above, appear as dark-coloured cellular nets, U-shapes, and doughnut-shapes on a lighter background. All tend to become elongated downslope and merge into striped patterns. The edges of nets and stripes are darker in colour and represent deeper soil. In southern England these patterns appear as crop marks. The lighter material filling the holes in the net is chalky drift whose upheaval appears to be due to the formation of self-accreting ice lenses within the soil matrix. We can see these in vertical views of the Isle of Thanet and the chalklands of East Anglia. They sometimes show segregation into stripes down slopes and become *rock glaciers* when flowing in a confined valley. Both stripes and rock glaciers result from the shattering of rock by frost weathering followed by gravitational movement. The shattering is thought to be due to cycles of wetting and drying combined with alternating sequences of freezing and thawing when minimum temperatures fall to below minus 5°C.

Stone pavements are accumulations of flat-lying cobbles and boulders fitted together into a mosaic, armouring and often completely mantling the surface. Continuous spreads on relatively level ground in high mountains or plateaus are called *blockfields*. They are thought to be due to block disintegration of underlying rock in situ from the combined action of ground saturation, the upward movement of stones through the periodic freezing of the water beneath them (called upfreezing), and the removal of the finest particles by meltwater. The stones are larger than the 'gravel path' scale of deflated desert pavements but the pattern is otherwise similar. Stone pavements differ from blockfields in that they overlie finer textured materials.

Stone polygons occur on level surfaces, *stone garlands* where the land slopes slightly and *stone stripes* where it slopes considerably. Each can merge into the next down a hillslope (**Figure. 12.10**). They are thought to be due to the different heaving effects of stones and fine materials when a mixture freezes. The stones tend to perch on the top of needles of ice which they preserve longer from melting. This leaves them on top of other materials after melting. When the surface freezes into polygons whose centres are higher than their edges, the stones migrate outwards to the polygon margins by gravity. The garlands and stripes appear to be caused by the downslope elongation of the same processes.

Avalanches result from snow melt. Tens of thousands occur in mountainous areas like Switzerland every year. The debris is roughly sorted but the largest blocks tend to fall furthest and to have debris tails at their distal ends. Where the avalanche is unconfined it develops a fan shape with the contained boulders generally aligned to the direction of flow.

Permafrost is hard and is treated like a rock for building foundations in subarctic areas, but in summer it can melt. This melting goes deeper under lakes, ponds, and trees, giving an uneven surface with boggy depressions somewhat resembling karst areas during a wet season. For this reason, it is sometimes called *thermokarst*.

CHAPTER 13

Alluvial landforms

Processes

Rivers and streams erode by backward extension of source gullies, scouring beds, and undercutting banks. The typical longitudinal profile of a river looks half-paraboloid (**Figure 13.1**). It begins with a small relatively steep stream. Where this is in a gorge or chasm and is floored by rapids and cascades, it is known as a *flume*. It then becomes wider and more nearly level towards the mouth. Under stable conditions a stream 'grades' its bed towards this form by eroding mainly in the upper reaches and depositing mainly in the lower. In some places, an eddy may rotate a stone so that it wears out a circular *pothole* in the underlying rock.

River deposits are of three main types: *bedforms* along the channel floor, *terraces* alongside it, and *alluvial fans* at its downstream end. The average size of particles deposited decreases from source to mouth. Where the flow is relatively constant stony bed material is usually arranged with the long axes of stones overlapping one another and leaning downstream - so-called *imbrication*. Where the flow is occasional, as in arid areas, the imbrication is less evident and the channel bed has

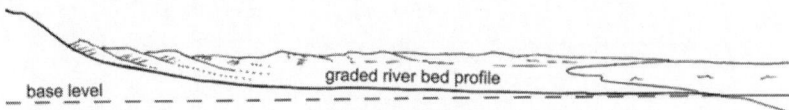

base level   graded river bed profile

Figure 13.1: The relation between base level and stream grades (after Holmes page 154).

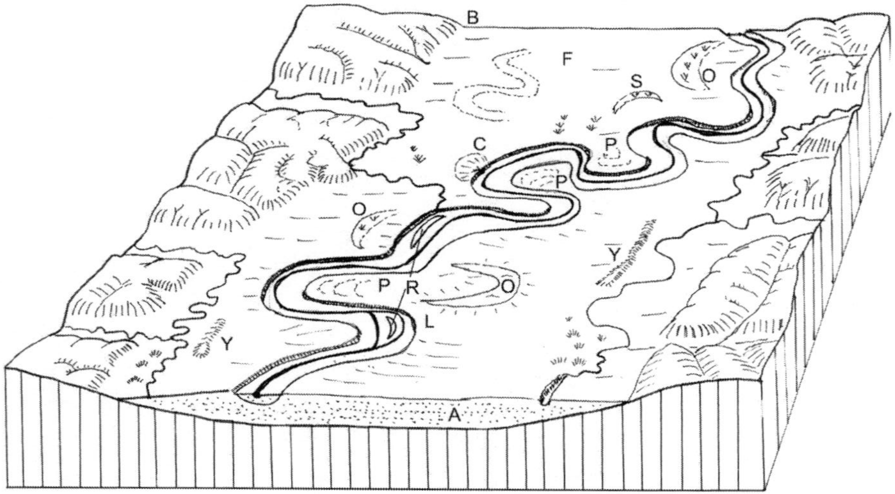

Figure 13.2: A mature meandering river: A: alluvium filling the valley floor; B: bluffs marking the edge of the meander plain; C: crevasse- splay deposits laid down when the river breaches its banks over a short distance; F: the whole flood plain of the river; L: levee - the higher ground beside the river caused by the deposition of sediment when it overflows in flood time; O: oxbow lake which is a crescent-shaped lake in an abandoned curve in the river course; P: point bar deposits which accumulate on the inside of a river meander curve; R: riffles which are depositional bars on the channel floor of the river; S: swales, which are troughs in the microrelief of the flood plain formed in an earlier depositional phase, usually juxtaposed with parallel bars; Y: yazoos are deposits of tributary streams which are prevented from joining the main river because of its flanking levees (after Strahler page 445).

elongated semi-parallel banks of coarser material and hollows of finer material aligned in the direction of flow.

Rivers have two most characteristic patterns of lateral movement: *meandering* (Figures 13.3 & 13.4) and *braiding* (Figure 13.6). Either or both can occur in any size of river. For a given slope, braided channels have a higher discharge than meandering channels. In general meandering occurs where a river has a constant gradient and braiding occurs below an abrupt lessening of it. Whether a particular channel will meander or braid can theoretically be determined by calculating what is called its *discriminant function*:

S - 0.012Q + 0.44, where S is the channel slope in metres per metre and Q is the bankfull discharge in cumecs (cubic metres per second). Where the value of the function is >0 the river will tend to meander, where <0 it will braid.

Meandering rivers

Rivers tend to flow in sinuous loop-like bends characterized by a *river-cliff* or *bluff* on the outside of the curve and a gently sloping surface on the inner side (**Figure 13.2**). This property of *sinuosity* is defined as the ratio of channel or *talweg* (channel mid-line) length to the length of the valley occupied by the channel. When sinuosity exceeds 1.5 the river is said to *meander*. A meandering river has a single channel that repeatedly swings first to one side and then the other of its mean flow direction. This causes a specific pattern of erosion and deposition. Well-known landforms result from this. It deepens and scours the outside of meander bends and leaves *point bar deposits* on the inside of the bends. These often separate moon-shaped hollows often containing water called *swales*.

Riffles are ripples in the loose material deposited in the river bed midway between the points of maximum meander curvature. Their height is 10-20% of the mean flow depth, and they have the form of anti-dunes with the side facing up-current steeper than that facing down-current.

Meandering rivers can deposit raised banks called *levees* of gravel and sand along their courses. Where one of these prevents the inflow from a tributary it is called a *yazoo*. From time to time the river bursts its banks and spills over the levee on to

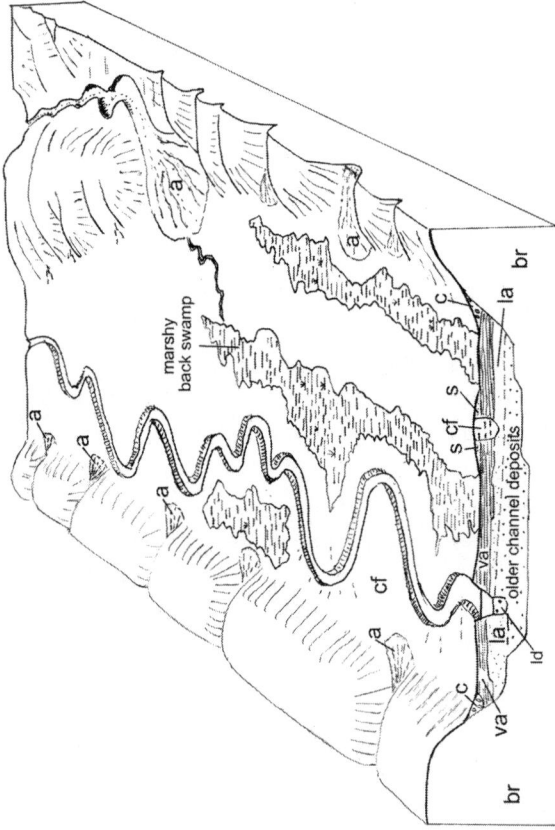

Figure 13.3: Types of alluvial deposits. c: colluvial deposits from the slopes along the valley sides; va: vertical accretion deposits from suspended-load materials laid down by the river; la: lateral accretion deposits laid down on the sides of channels where bed-load material are being moved towards the inner side of meanders; s: splays of material deposited along the sides of a former channel; ld: lag deposits which are of coarse material left behind on the bed of a stream; cf: channel fill formed of large bed-load materials; a: alluvial fans deposited by streams which lose gradient where they meet the flood plain; br: bedrock base (after Thornbury page 171).

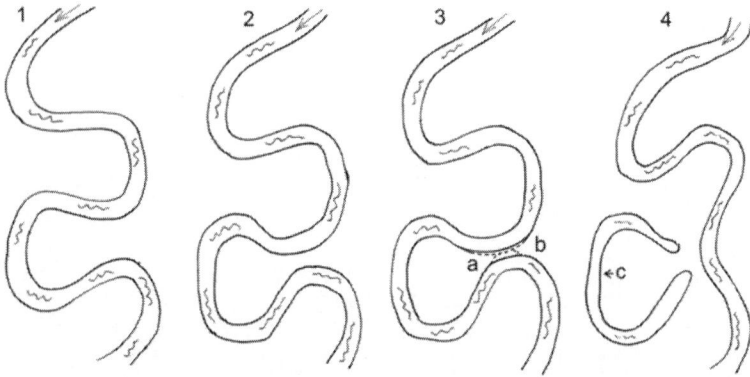

Figure 13.4: The development of meanders showing the creation of an ox-bow by the cut-off of a loop (after Holmes page 165). Note that in stage 3 the river rejoins its own channel (dashed line) by undercutting the intervening neck at river level and leaving a natural bridge (a-b) above it (see also Fig 13.5). In phase 4 the oxbow lake (c) is left isolated.

the flood plain. This usually occurs at the outside of meander bends. In doing so it lays further deposits, called *crevasse-splay deposits* immediately beyond the point where the overflowing river breaches it banks. The flood then deposits the finer materials over the extent of its flood plain. Particle sizes decrease to a minimum in lowest *backswamp* areas (**Figure 13.3**) where the slack water deposits clays. This 'fining' of particle sizes away from the river is often important to land uses.

Where the levee has risen to an unstable height, the river abandons it and repeats the process in another place. By this means some rivers have made considerable lateral migrations. Their flood plains are therefore like palimpsests of past meander courses appearing as a changing *scroll pattern* when viewed from above. They show the past circumstances of erosion and alluviation in the valley where each succeeding stage part buries and part exposes the preceding one. This presents a complex problem of interpretation similar to that of reading an ancient document which has been erased and written over a number of times.

Where meanders become especially developed two neighbours can intersect. This causes the stream to change course "short circuiting" the meander. The cut-off meander usually becomes

Figure 13.5: Rainbow Bridge, Bridge Creek, Utah. This is a natural bridge formed by the backwearing of two consecutive incised meanders which have undercut their separating neck from both sides (drawn by Julian Mitchell after Holmes Plate 44b).

plugged by sand at each end, leaving a central curved depression which fills with water. Once isolated the cut-off receives only fine materials which tend to give it an impermeable floor. It becomes an *ox-bow lake*, resembling the Australian *billabong* except that the latter can also include ephemeral streams (**Figure 13.4**).

When two incised meanders undercut the narrow neck between them each can form a cave, especially when the rocks at river level are weak.. Eventually the two caves meet and the two streams join through the perforation, leaving a natural bridge and an oxbow lake **(Figs 13.4 & 13.5)**.

The longer the time since a river channel was abandoned the more obscured the scroll pattern becomes. Three zones of diminishing articulation of the pattern are sometimes recognizable in the flood plains of large rivers: an *active flood plain* where the river still flows; a *meander flood plain* where it occasionally floods and the scroll pattern is still clearly expressed, and a *cover flood plain* where it has ceased to flood and the scroll pattern has become largely obscured.

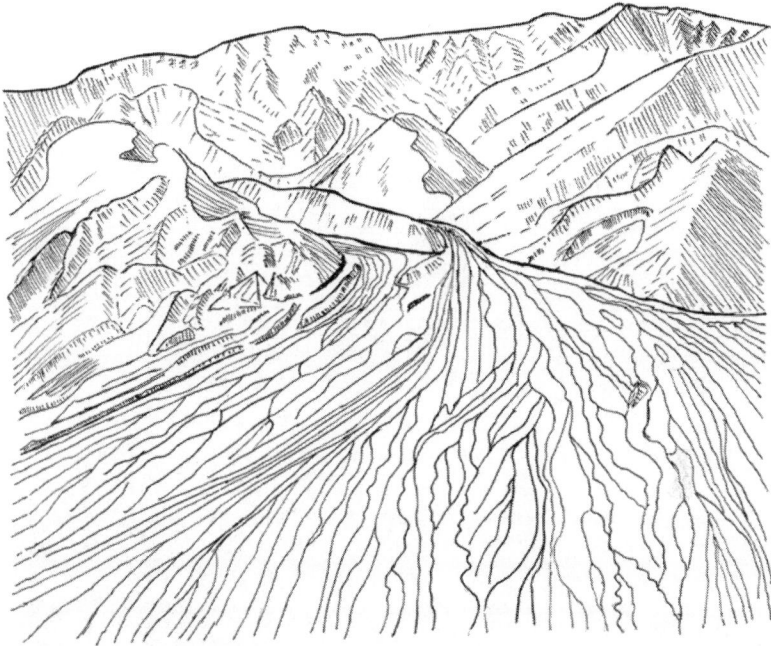

Figure 13.6: Stream braiding on a great alluvial fan in Death Valley, Idaho (from photograph in Strahler page 452).

Braided streams

Braiding occurs where the sediment-charged water experi ences a sudden decrease in gradient or discharge. The excessive sediment load causes the channels to bifurcate, branch, and rejoin irregularly. It thus forms a network of interconnected converg- ing and diverging channels resembling the strands of a braid. The unconfined flow of runoff water is sometimes called *sheet- wash*. This differs from confined flow only in degree since much unconfined flow is in quasi-parallel surface channels. It is most effective in deposition when heavily charged with sediment. This occurs most markedly where it carries materials eroded from a relatively unvegetated catchment and where deposition is induced by a rapid diminution in gradient. The characteris- tic location is at the mouth of a mountain stream on to a plain, where it forms an *alluvial fan* (**Figure 13.6**), deriving this name from the radiating pattern of channels. Stones and gravel are concentrated on the higher strips (**Figure 13.7**), silts and clays in

Figure 13.7: Section across upper part of an alluvial fan, Oman. Such fans are formed when a mountain torrent debouches on to the plain. Its energetic carrying of loose materials is suddenly arrested, causing it to deposit them in layers, the coarsest materials first and then fining upwards. Note the stratification in the finer materials.

hollows occupied by ponds at times of lowest flow. Sands generally occupy the intermediate topographic situations.

Alluvial fans and deltas

The smaller the alluvial fan the steeper the slope. The profile along the direction of flow is generally concave upwards, but is sometimes straight over the whole or a substantial part of it. The crosswise profile generally has rises called *interfluves* punctuated by channels. The distal end is usually lobate when viewed in plan. The channels migrate laterally as they progressively build and abandon levees. As the fan grows, its main locus of deposition extends outwards, tending to cannibalize the upper parts to feed the lower. It is sometimes possible to unravel the stages of this. If the base level of erosion rises, erosion lessens and channels silt up. If it falls, erosion increases, and drainage channels become incised, leaving remnants of the former surface upstanding. Sometimes several levels of progressively decreasing altitude indicate successive stages of downcutting, as on the southern slopes of the High Atlas in Morocco.

Viewed in plan, channels across alluvial fans show three zones of changing directionality. Near to the apex they point more or less directly out from the source. At the distal end, they are generally oriented to the direction of the drainage out-

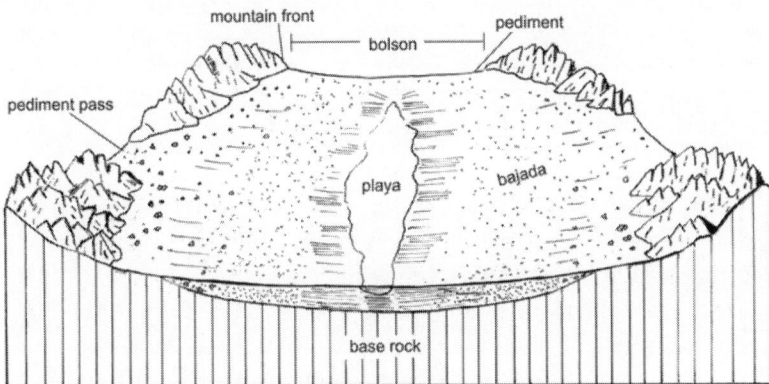

Figure 13.8: Typical assemblage of desert landforms. The top of the slope is bare pediment. Below this the material fines from boulders to clay with the playa highly salty.

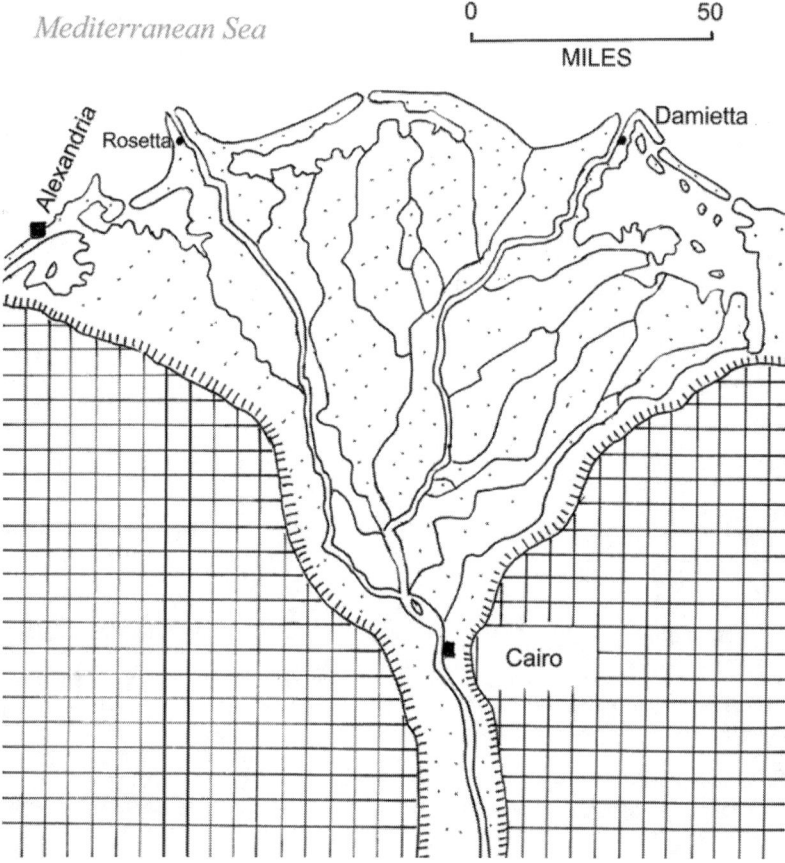

Figure 13.9: The Nile Delta (after Lobeck, 1939, page 234). This is the alluvial area deposited by the Nile which terminates at the Mediterranean coast. Note its triangular shape with apex at Cairo, reproducing that of the capital letter delta in the Greek alphabet from which the name derives.

fall. In between, they anastomose as they change direction and carrying capacity with the decreasing gradient.

Along a mountain front, a series of adjacent fans may coalesce into a single *alluvial apron* which in arid areas is called a *bajada*. Below this there is often a wide area of unchannelled alluvial outwash called by the French term *bord*. The lowest ground is usually occupied by a lake in humid areas, or a seasonally flooded salt-charged *playa* in arid areas. In arid areas the whole basin between neighbouring mountains is called a *bolson*. In parts of southwestern USA the rapid crumbling of material detached from the mountain front has left the upper part of the eroded slope as a *pediment* almost bare of detritus (**Figure 13.8**).

A delta occurs where a silt-laden river loses gradient and deposits its sediment at the coast of the sea or a lake. The name derives from the plan view of the Nile delta which resembles this Greek letter (**Figure 13.9**). It is essentially an alluvial fan, though with differences. The deposited material is generally finer in texture after its relatively long river journey. Gradients are lower and the presence of the sea makes the delta everywhere subject to a shallow water table. There is often standing water away from natural levees. Distal areas are affected by tides, and stream mouths diverted and sometimes blocked by longshore drifting.

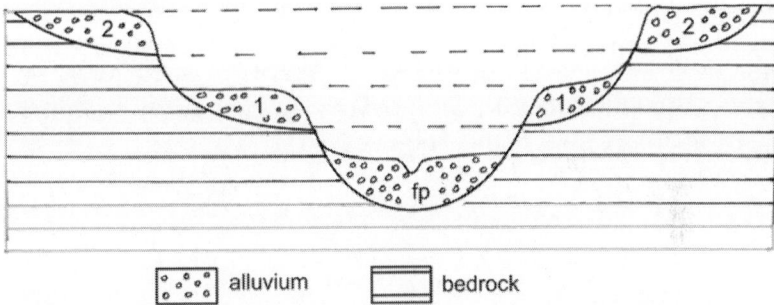

Figure 13.10: Cross section of a valley showing river terraces: fp flood plain; 1 first above floodplain; 2 second above flood plain (after Gorshkov and Yakshova page 143).

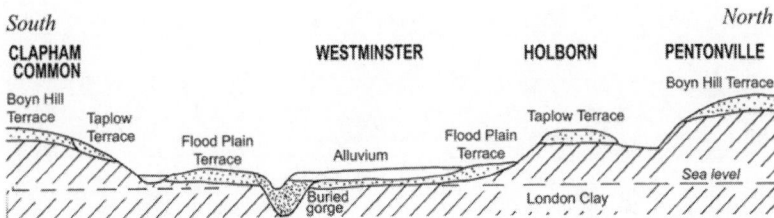

Figure 13.11: Section across London showing alluvial terraces (after Holmes page 196).

Deltas vary in overall plan according to the configuration of the coast on which they emerge. Where they have radial distributaries and a rounded outer coast, as in the Nile or Rhone deltas, they are called *arcuate*. Where the delta has grown out finger-like at the end of each distributary, as in the Mississippi delta, it is called *bird-foot*. Where a river emerges from a fairly straight shore line along which wave attack is vigorous, the sediment is spread along the shore in both directions from the river mouth, making a *cuspate delta*, as at the mouth of the Tiber. Where the river empties into a long narrow estuary, the delta is confined by its shape to form an *estuarine delta*, as for the Seine.

Terraces

A river intermittently downcutting its bed leaves remnants of former levels called *terraces* (**Figure 13.10**). At the time of formation these were levels of alluviation, but when the river cut its bed deeper they were left as elevated residual flats. A major river will generally display a 'flight' of such terraces showing the stages of its history. The Thames in the London area, for instance shows this (**Figure 13.11**).

Chapter 14

Mudflats

Mudflats develop in basin sites at the lowest position in the landscape and are the flattest and most level of all land-forms. They originate as marshes, lake beds, or areas of slack water deposition off coasts. Water draining into them has already dropped all its coarser particles. They receive only the fine silts and clays, which therefore form the basin floor. The impermeability of these materials causes them to hold up lakes which are permanent in humid regions, seasonal or ephemeral in arid.

In humid regions mudflats tend to support abundant vegetation and so accumulate organic matter. This makes them fertile when drained. Some of the richest soils in Britain, notably in the Fens and south Lancashire, are of this type. In arid regions mudflats give level shallow basins called *playas* containing seasonal or ephemeral lakes. Evaporating water concentrates salts, zoning them inwards in order of increasing solubility in the general sequence: magnesium-calcium-sodium and carbonate-sulphate-chloride. Accordingly, sodium chloride dominates the centre of playas although larger amounts of gypsum may be contained in the sediments. They have three types of surface: polygonally cracked clays (**Figure 14.1**), soft puffy surfaces where ground water brings some salts to the surface (**Figure 14.2**), and salt crusts where these are overwhelmingly dominant. Spring mounds (**Figure 14.3**) occur where artesian conditions bring lime and gypsum to the surface under pressure and evaporation accumulates them in eminences above the surface.

Mudflat pavements crack into *takyr* when dry (**Figure 14.4**). This is generally more marked in clays (particle size less than 0.002 mm) than in silts (particle size 0.002-.05 mm). The cracks first appear across the middle of the longest dimension because

Figure 14.1: Oblique view of cracking clay with major cracks about 60 cm apart, Iceland (after Schmid page 59).

this represents the direction of maximum tension. They then proceed to subdivide the rest of the surface on the same basis until there is an approximately homogenous pavement whose fineness of subdivision reflects the maximum degree of drying achieved.

Figure 14.2: Surface with evaporating salts in an arid area somewhat resembling rising dough under a thin broken crust, Jafr depression, Jordan. The reason for this appearance is that when a heavily saline mixture of clay and salt evaporates the salts crystallize and strongly flocculate the clay.

Clay mineral behaviour

Clay particles are platy in form. Their physical behaviour depends largely on the extent to which they flocculate (cling together in clusters). This tendency to flocculate increases with the total amount of soluble salts in the soil water and with the ratio between the positive metallic ions, especially calcium and sodium, on the total of ion-holding particle surfaces, called the exchange complex. Although calcium is usually dominant, sodium becomes so, for instance, on lands vacated by seawater or in dischargeless basins in arid areas where salts are concentrated at the surface by evaporation, and only halophytic (salt-

Figure 14.3: Spring mound about 2 metres high, Danakil depression, Ethiopia (from a photograph).

loving) plants can live. If the salt is then washed out by further flooding of the surface, residual sodium ions can still dominate the soil's exchange complex. The dominance of sodium ions not only deflocculates and impermeablizes the clay but also reduces its fertility by increasing alkalinity and depressing the availability of calcium to plant roots.

Because of their cohesiveness, clays exist in three distinct states, depending on the water content. First, in the dry season in the tropics, or in summer in temperate latitudes, they are hard, often cracked, difficult to cultivate, but easy to traverse. Secondly, in the wet season in the tropics and under the reduced evaporation in autumn in temperate regions the clay softens. This is the condition making it most workable for agriculture. Adding more water causes it to passes its *plastic*

Figure 14.4: Takyr surface. Note that the larger cracks appear first and are then subdivided. The board is about 30cm across, California, USA.

limit. The surface then becomes malleable and sticky, and dotted with ponds and puddles. This makes it heavy to cultivate and slow and laborious to traverse, until the seasonal return of drier conditions. Thirdly, under very wet conditions, clays will exceed their *liquid limit.* They then flow as a viscous clay-water mixture. This occurs seasonally on some tropical clay plains, where the whole surface moves in fluid surges. These often have lobate edges across the direction of flow, which can sometimes be identified from curved vegetation patterns.

Landscape effects of clay mineral types

The landscape effects of clays also depend on the type of clay mineral composing them. There are three main types, known generally by their commonest representatives as *kaolinite, montmorillonite,* and *illite* (**Figure 14.5**).

Kaolinite micelles (individual particles) are relatively large (0.1-5 microns). They are built of double layers of molecules

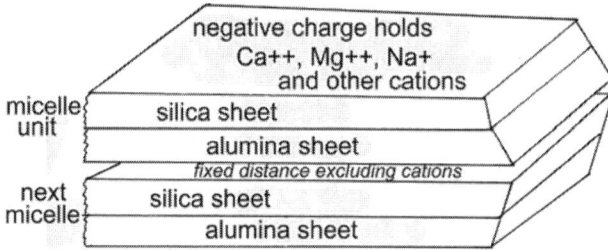

Figure 14.5a

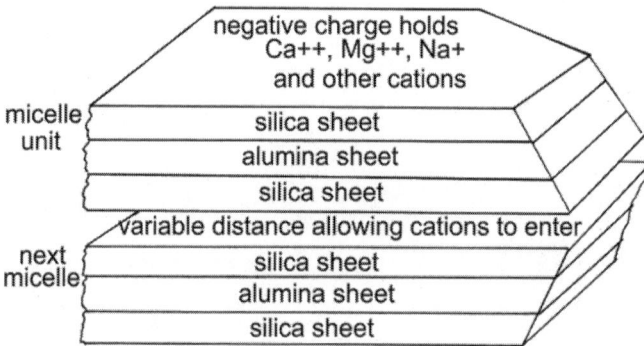

Figure 14.5b

Figure 14.5: Diagrammatic representation of clay micelles with their sheet-like structure: a) kaolinite, b) montmorillonite. Illite resembles montmorillonite except that potassium ions add further structural connection between the sheets, increasing fixation between micelles. Each particle is less than 0.005mm (5 microns) across.

consisting of a sheet of silica bonded to a sheet of alumina by oxygen atoms. This is called a fixed-lattice clay. The double layers are too close together for water or metallic ions to penetrate between them from the soil. They thus cannot expand much on wetting nor contract and crack much on drying.

Montmorillonite micelles, by contrast, are much smaller (c0.01 microns) with a 2:1 expanding lattice structure made up of two sheets of silica sandwiching one of alumina. The inter-sheet spacing is variable and wide enough to allow water and cations to enter from the soil, so has been called an expanding-lattice clay. Because of these characteristics such clays have a much greater overall surface area for any given volume of material. This makes them more plastic and cohesive. Because they are able to absorb more water, they swell much more on wetting, and shrink and crack much more on drying.

Illite is often associated with montmorillonite. It likewise has a 2:1 structure but differs in having larger particles (0.1-2 microns), some silicon atoms replaced by aluminium atoms, and the negative charges largely satisfied by potassium ions. Illite is thus intermediate in character between kaolinite and montmorillonite in its susceptibility to expansion and cracking.

The formation of clay minerals depends on the environment. Weathering from parent rock is accelerated by high temperatures. Montmorillonite tends to form in alkaline environments where there is an abundance of metallic cations and restricted drainage whereas kaolinite tends to form under acid conditions where the cation supply is limited. Illites are intermediate and especially common where the soil solution is rich in potassium, as on unleached rocks of marine origin. For these reasons, montmorillonite and illite tend to be relatively more frequent in arid areas, kaolinite in humid areas.

The consequences in the landscape are significant. Because kaolinitic clays are less water-absorbent and crack less, their slopes are less vulnerable to erosion. This makes them better than montmorillonite at sustaining structures and more useful for building materials and ceramics.

Montmorillonitic clay soils are common on plains in the tropical savannas. They expand and flow in the wet season, and sometimes crack so widely and deeply in the dry season that trees and posts are skewed and considerable loose materials from the surface fall down the cracks, to be trapped and churned when the wet season returns. Buildings whose foundations do not go below the zone of cracking, or which are not founded on stable rafts, can suffer severe cracking.

The fertility of clay soils depends on the chemistry of the surfaces of the clay particles. Montmorillonite particles, having a larger overall surface area and higher negative charge are also able to adsorb many more positive metallic cations than kaolinites. This, combined with their greater water holding capacity and permeability when flocculated, makes them potentially more fertile than kaolinites. Illites are again intermediate.

CHAPTER 15

Organic terrain

Organic terrain is land whose surface is composed of vegetable matter. This most often consists of peat formed under waterlogged conditions (**Figure 15.1**). It develops in two stages: mineralization to break plant litter down into simple organic compounds, and humification to build these into complex colloidal molecules. Where the waterlogging has been total and lasting, the plant materials change to a pale yellow spongy mass. Where the process has been interrupted by drier periods allowing more complete mineralization and humification, the peat is dense, dark, and better humified. Because of their waterlogged anaerobic (oxygen-free) state, peats provide

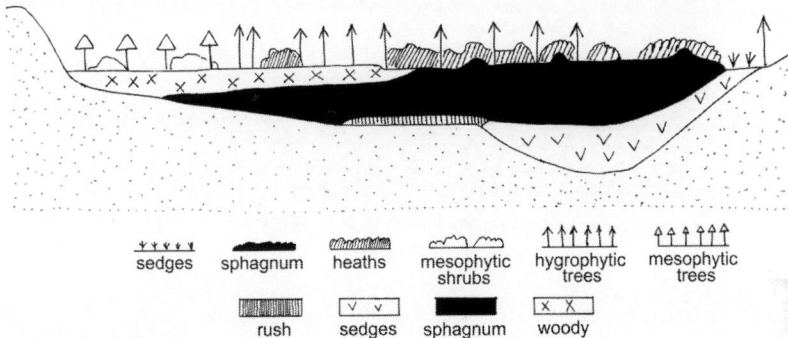

sedges   sphagnum   heaths   mesophytic shrubs   hygrophytic trees   mesophytic trees

rush   sedges   sphagnum   woody

Figure 15.1: Typical mature bog developed around a shallow pond or lake of glacial origin in a old continental climate such as that of Canada and northern Europe. First to come are the mesophytic trees and shrubs, then the hygrophtyic trees on the surrounding land. Then gradually the open water is replaced by an organic mass, here of sphagnum moss, with sedges and the rushes at the water's edge, (illustration from the (Laurentian Shield area, Canada, after Strahler page 336).

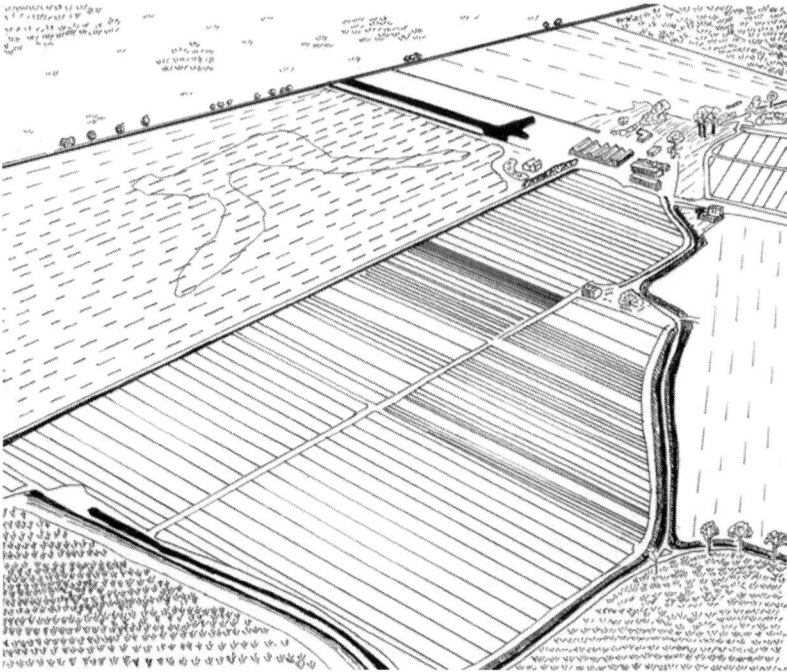

Figure 15.2: Reclaimed fenland near Spalding Lincolnshire. Note the drainage ditches shown in black around fields. The nearer lined area is growing tulips. The wavy line on the upper field relates to a former drainage pattern (after Stonehouse page 114).

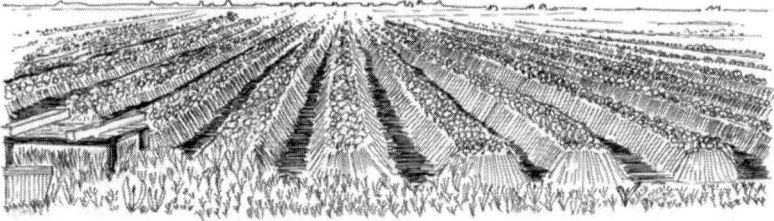

Figure 15.3: Soil, celery, and sky in fenland (after Miller Figure.70).

a medium which preserves woody and other hard materials, often for centuries.

The chemical character, sometimes called the trophic state, of peat is defined by its acidity. *Eutrophic peat* is formed where lime and bases are available and the soil relatively non-acid. It tends to have a rich vegetation with a lush carpet of sedges and mosses.

Once drained, it is fertile, as in the fens of England and Holland (Figs 15.2 & 15.3). *Oligotrophic peat*, by contrast, is formed under acid nutrient-poor conditions. The vegetation is mainly mosses, acid-loving grasses, and heaths and it has low intrinsic fertility. *Mesotrophic peats* are intermediate between these two.

Peat occurs mainly in three types of geographical environment: *tundra* in high latitudes, areas with high rainfall and impeded drainage in mid-latitudes, and basin sites in humid climates generally. Tundra is that zone of the earth's surface underlain by permafrost. This prohibits soil drainage. The surface therefore alternates between frozen conditions in winter and waterlogged conditions in summer. It can only support a low tussocky vegetation whose trophic state depends on the 'base status' or 'base saturation' of the soil (the extent to which its absorption complex is saturated with exchangeable cations other than hydrogen and aluminum). There are occasional trees in places where the soil is better drained. There is a dense mat of vegetation over peaty soil but disturbances to this are slow to recover and the scars often survive for years. The country is monotonous to view and very hard to traverse. But at times it can have great beauty, notably after the snows have melted and there is a flush of spring flowers.

Hills and areas with shallow rocks in cool humid climates sometimes develop a covering of *blanket peat* (also called *mountain peat* and *blanket bog*). This is due to an excess of soil water in situations of high rainfall with low evaporation and restricted drainage. Since it generally occurs on acidic siliceous rocks, it is treeless and supports acid-loving vegetation dominated by mosses, in Britain mainly *Sphagnum* species. The result is to make the peat oligotrophic. This situation is widespread in Highland Britain (**Figure 15.4**), where there is an additional reason for the poor drainage. Many of the hills were originally forested. The trees enriched the soils with plant nutrients brought up from the deep subsoil by their roots and dropped in the leaf litter. But then, it is believed, early settlers deforested these hills in order to cultivate or graze them. This broke the fertility cycle and caused the soils to acidify, so that iron was leached and re-deposited as an impermeable iron pan in the subsoil. This pan has hindered subsequent through drainage and contributed to the surface waterlogging, in turn causing the blanket peat.

Figure 15.4: An oblique view of a large raised bog showing a concentric arrangement of open water (black) or filled with floating Sphagnum (blank with grass symbols) with sandy areas (dotted), The largest ponds in the foreground are around 100 metres long, Silver Flowe, Kirkcudbrightshire, Scotland (after St Joseph plate 36).

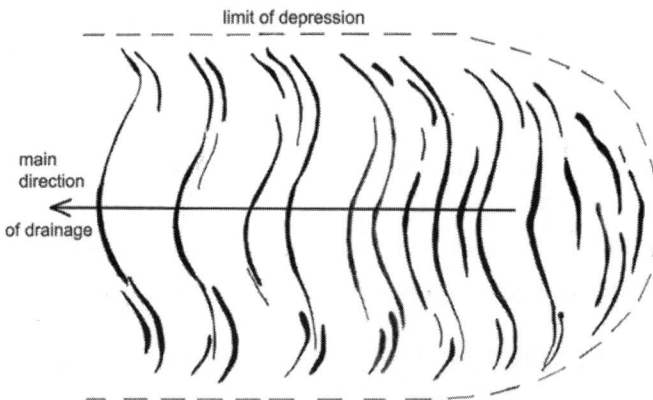

Figure 15.5: Plan view of typical orientation of string-bogs, as shown by the wavy lines, in a wide depression on Adelaide Peninsula, Canada. These are boggy depressions, usually more than 100 metres in length, in gently sloping outwash plains where the material moves downslope in lobate form, bending around resistant obstacles such as knolls, better drained sites, or bedrock (after Henoch page 337).

Blanket peat can vary in thickness from a few centimetres up to several metres. It is sometimes cut by erosion channels into isolated columns called *groughs*. When masses of peat on hillsides become detached and move downslope, they are known as *hags*. In some areas bogs have surfaces showing patterns, festoons, and nets. In the Canadian Arctic these patterns are best developed in boggy depressions on gently sloping outwash plains, where they are called *string-bogs*. Their ridges bend where they meet broadside knolls, better drained sites, or bedrock. In areas of permafrost they tend to concentrate around lakes (**Figure 15.5**).

Waterlogged basin sites develop *basin peat*. The character of the peat derives from the nature of the vegetation it supports, which in turn reflects the chemistry of its contributing river catchment. *Marsh* refers to waterlogged land with a soil that has a siliceous (quartz-rich) mineral base, *fen* to a similar situation on lime-rich soil and a vegetation which includes some trees, and *carr* to a similar situation with many trees.

Swamp differs from all these in that even in summer the water level is above the ground surface and so the vegetation is always in standing water. It is usually dominated by reeds, sedges, and rushes. If the water level falls it becomes marsh or fen.

Marsh is common in estuaries and coastal lagoons, and around the margins of lakes such those in the Lake District. The English Fens are a good illustration of an area of eutrophic basin peat which has been drained (Figs 15.2 & 15.3). Their original state can be seen at Wicken Fen which also includes areas of *carr* where the soils are not too acid or deficient in mineral nutrients and support some trees such as alder and willow. These patterns can still be discerned in drained fenland. Bog occurs widely in the northern Pennines on the borders of Yorkshire and Cumbria, and swamp around the edges of ponds and lakes everywhere where vegetation is growing on mossy ridges separating patches of standing water or sedge meadow.

When peat is buried beneath clay and sands it compacts. As the overhead pressure increases and water and gases continue to be driven off, the carbon content of the residue is continually increased until it changes into a variety of coal. The first form is lignite but the continuation of the process alters it first into bituminous coal and ultimately into anthracite when unusually high pressure due to deep burial combined with an increase in temperature has driven off nearly all the volatile constituents. Where coal is exposed it forms a rock and can be mined as opencast.

CHAPTER 16

Marine sediments

Wave deposition

Waves on a rocky coast operate like a horizontal buzz-saw, cutting a vertical cliff and removing the fallen detritus to leave a wave-cut bench (**Figure 16.1**, see also **Figures 2.11 & 7.11**). The breaking wave digs a slight hollow in beach material. It then moves up the slope of the beach in a movement called *swash*. This is strong enough to move materials of substantial size such as pebbles and cobbles. Its alternation with *backwash* accounts for the sorting of beach materials with the coarsest shingle at the top of the beach and the finest clays furthest offshore. The beach material acts as a buffer which dissipates energy, so that over the long term the swash tends to be more effective than the backwash in moving it. Backwash, however, gives the materials their final orientation. Evidence of this is imbrication with pebbles leaning against each other towards the sea. (**Figure 16.2**). If sand is arriving at a section of the beach more rapidly than it can be carried away, the beach is widened and built seawards, a change called *progradation*. If, on the other hand, it is leaving more rapidly than it is brought in, the beach is narrowed and the shoreline moves landward, a change called *retrogradation*.

The limit of the swash at high tide is marked by a linear accumulation of shingle with an upstanding crest called a *beach ridge* or *berm* (**Figure 16.3**). Although berms are formed at most high tides, only those formed at high spring tides or by storm tides survive later destruction. When the beach ridge is formed during a stationary phase in an outward-growing delta it is called a *chenier*, and a series a *chenier plain*. Examples can be seen in the Mississippi delta.

157

Figure 16.1: Birling Gap, Sussex: chalk cliff above wave-cut bench with flints. Note how the sea has operated like a buzz saw cutting a horizontal bench in the chalk and leaving a vertical cliff above high tide. The chalk contained bulbous nodular flints which were hard and insoluble. The removal of the chalk by erosion has left these as a lag deposit on the bench.

The grain sizes of beach materials become coarser as beaches become steeper or narrower, especially when they are both. Also, exposed beaches tend to have coarser sand than protected beaches. This means that if we know three of the four factors: particle size, beach width, beach gradient, and degree of exposure, the fourth can be predicted with fair accuracy.

On regular coastlines outside protected bays, wave crests usually approach a shoreline obliquely. They are arrested and drawn more directly towards it by the shallowing of the water. This is called refraction. The wave resulting from the oblique component gives the energy to cause a longshore drift of materials (**Figure 16.4**). In Britain over 75% of these materials are supplied from rivers, the rest from cliffs. The process leads to the formation of elongated deposits of sand, shingle or mud called *bars*, of which there are a number of types.

Longshore currents sweep debris from headlands into the deeper water offshore where, losing velocity, they dump it in embankments. As long as the far ends of these are free they

Figure 16.2: The retreat of the tide piles beach pebbles so that they lean against each other down towards the sea in a manner called imbrication, at Port Elizabeth, South Africa.

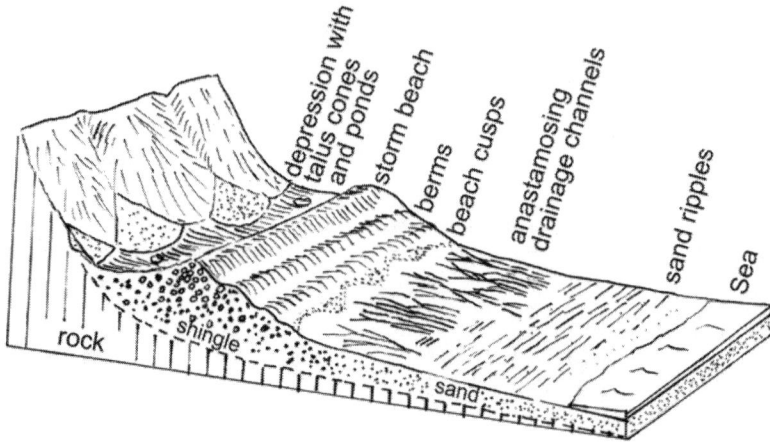

Figure 16.3: Idealized sand & shingle beach (after Small page 453). Note how high tides pile storm beaches and cause the decrease of particle sizes from the top of the beach towards the sea. The sandy beach experiences deflation and some dune formation, its upper part shows channelling by runoff water and the lower part shows ripples in the fine sand caused by retreating sea water.

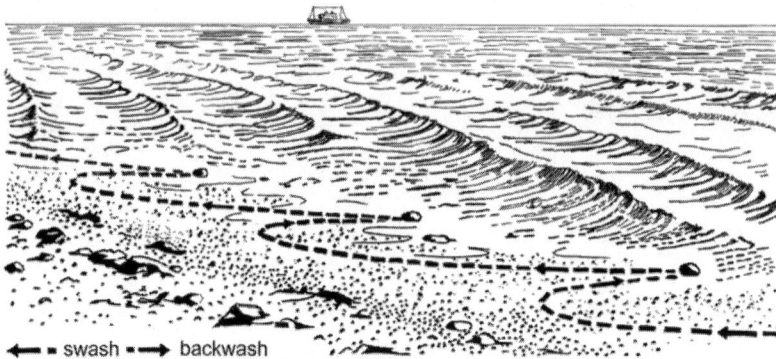

◄── = swash ──► backwash

Figure 16.4: Lateral movement of pebbles along a coast facing an oblique wave approach (after Shepherd page 43). This leads to a gradual migration of beach materials in the direction of tidal movement.

are called *spits* or *sandspits* (**Figures 16.5, 16.6 & 16.7**). These may change alignments to give characteristic angular forms at points where a different direction of wave action becomes operative. They are then called *recurved spits*, or if the direction of

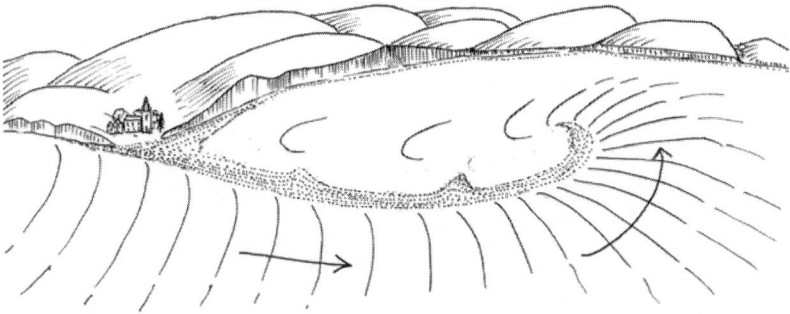

Figure 16.5: The formation of a spit by longshore drifting. Note the 3 horns representing consecutive storm beaches (after Shepherd page 49).

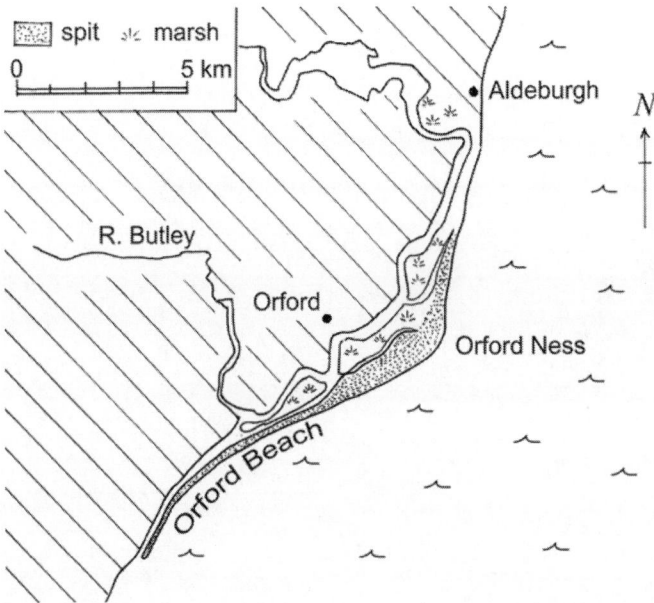

Figure 16.6: Orford Ness, Suffolk. The long spit of Orford Beach (dotted) was created by the southward-moving sea current. Note that the outflow from the rivers prevents the spit from joining the land.

movement is almost reversed, *hooks*. When breached they leave barrier islands.

Longshore bars are sand ridges situated in the intertidal zone and roughly parallel to the shoreline (see **Figure 16.6**). They may be submerged at high tide when waves break over them, but are exposed at low tide. They are often joined to the

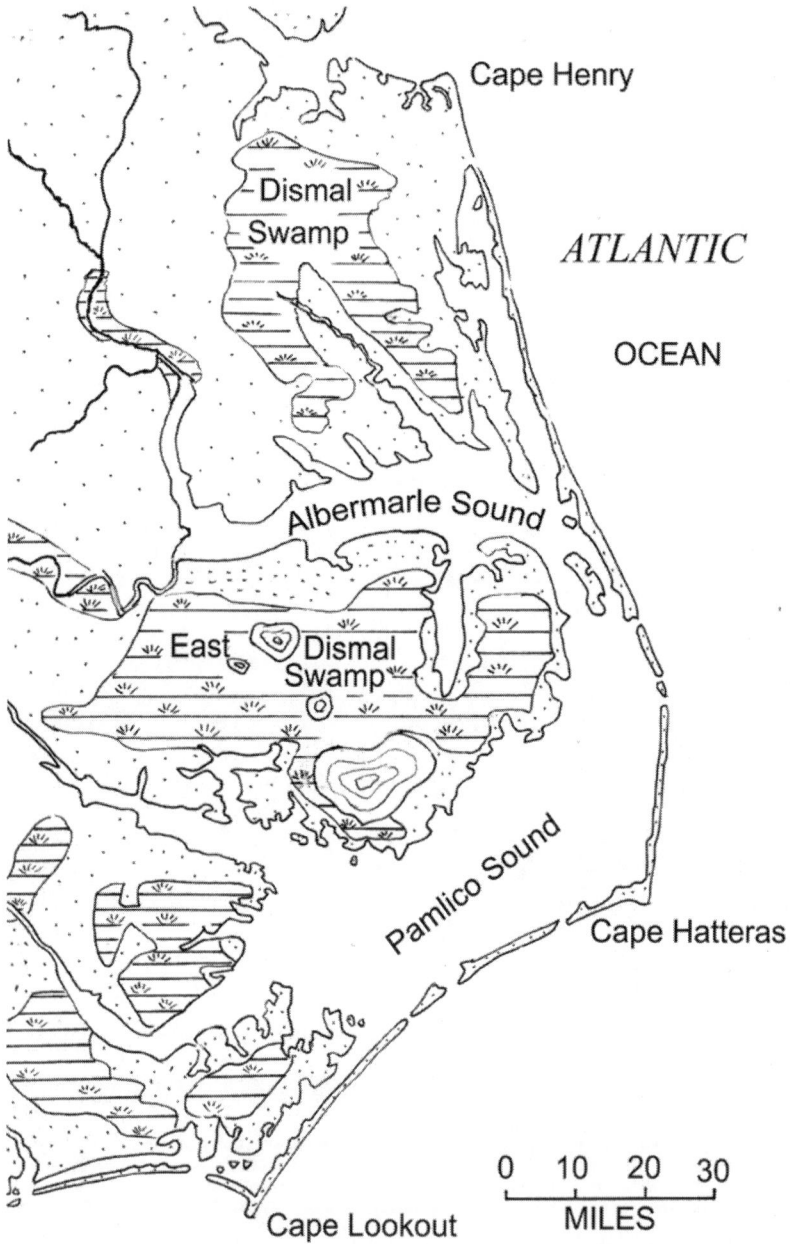

Figure 16.7: Barrier beaches and swamps along the coast of North Carolina. Note the rapid diminution of the width of sea channels as they go inland (after Holmes page 305).

shoreline at one end and separated from each other and the shore by troughs. The seaward (stoss) slope is gentle but the landward (lee) slope is usually steeper - at the angle of rest of the material. The orientations of surface ripples normally show that the main directions of water movement are over the bar and down the trough. Longshore bars which are not connected to the coast are called *offshore bars*. Offshore bars with crests above tidal level are called *barrier beaches* (**Figure 16.7**), and if high enough for dunes to grow on them, *barrier islands*. These are often separated from the coast by a lagoon or sound in which coastal marshes, or in the tropics mangrove swamps, may appear. The German term *haff* is used to describe the shallow coastal lagoon resulting from the growth of a spit across a bay or river mouth. A bar which encloses a bay is called a *baymouth bar*, and one which joins an island to the coast is called a *tombolo* (**Figure 16.8**). Some more complex cuspate forms derive from an alternation of direction of beach ridge formation. The foreland of Dungeness, for instance, consists of marshland behind beach ridges built up by storm waves both from the southwest along the English Channel and from the east which is open to the southern North Sea (**Figure 16.9**). Some beach cusps may reflect a combination of causes where there are both alternating onshore current directions and rip currents.

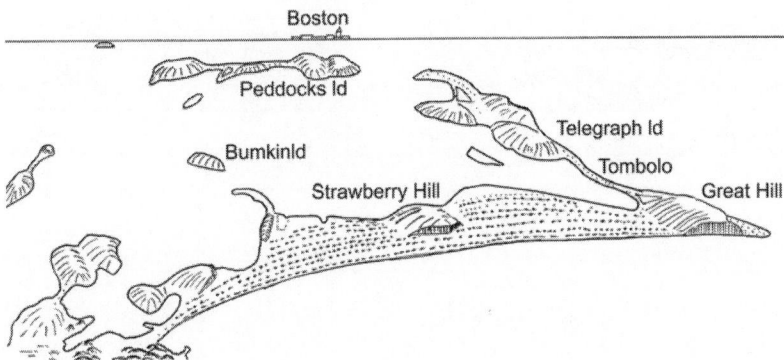

Figure 16.8: The complex tombolo of Nantasket beach, Massachusetts, formed by longshore drift from the south (left of picture). Note how it has joined several islands. The view looks westwards towards Boston (after Lobeck, 1939, page 352).

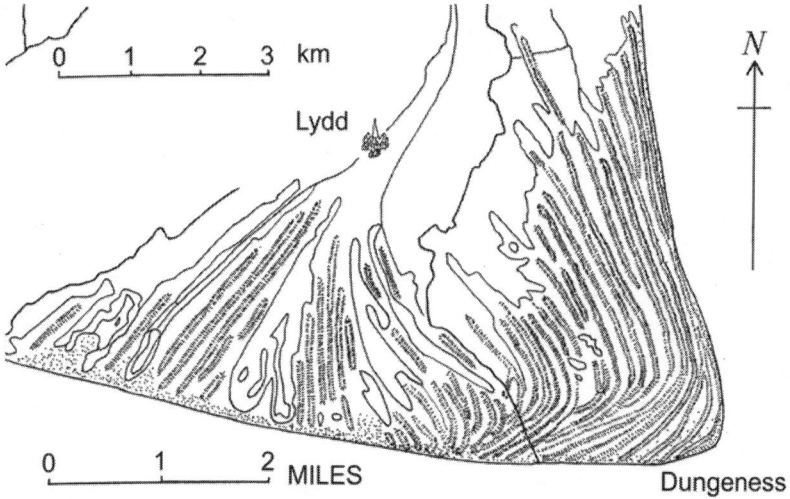

Figure 16.9: Dungeness Foreland showing beach ridges formed by the interaction of sea currents from west and east (after Strahler page 531). It is now reclaimed land.

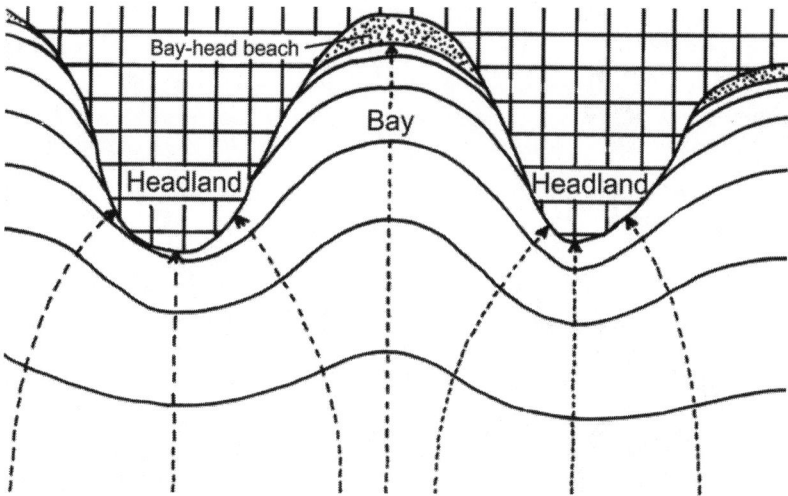

Figure 16.10: An indented coast (squared pattern) exposed to direct sea attack. The waves (wavy lines with directions shown by dashed lines with arrows) are refracted. Note how energy concentrates against head-lands and diffuses around bays. This tends to develop the coast towards a stable linear form with a continuous beach (after Holmes page 285).

Occasionally, bars are parallel and extend outwards from the coast. They may sometimes be due to waves losing momentum and thus dropping sediment as they move over slight prominences on the sea bed. I they are regular and equidistant they are probably due to unidirectional longshore currents.

On an indented coast the headlands impose the first shallowing, the bays only later. Such headlands refract the wave front as shown on **Figure 16.10**. They face high-energy attack and mainly suffer erosion. The bays, by contrast, are low-energy zones where wave force diffuses around the bay and depositional forms predominate. The debris deposited around the inland end of the bay forms a *bay-head beach*. Only very destructive storm waves can affect its backshore (the zone above normal high tide). Sediment movement here is mainly up and down the beach.

When strong winds cross a beach they dry it and entrain the fine particles, moving them inland into dunes. Imposing dune complexes are often formed in this way. Whenever the beaches are in areas where there is some rainfall the dunes become stabilized by vegetation.

Tidal Action

The tide rises and falls on average every 12 hours and 26 minutes. It is caused by the moon, and to a lesser extent the sun, pulling out bulges of water on the earth by gravity. Tides have a maximum vertical range when these two are in line with the earth. These are called 'spring tides'. When they are at a right angle as seen from the earth tides are lower and are called 'neap tides'.

The amount of tidal rise and fall can vary widely around a coast. It is minimum on open coasts and reaches a maximum on long narrowing inlets. In Britain, although it is mainly between 3 and 7 metres, it can vary from less than 2 metres on exposed coasts, such as east Norfolk or Cardigan Bay, to 12.3 metres at the head of the Severn estuary with the famous 'Severn bore'. The tide in the inner end of the Bay of Fundy in Canada has a record range of about 19 metres.

The height of the tide governs the width of the beach. Where the height range is less than 2 metres the approaching sea currents are concentrated over a very narrow height range. This makes them more effective in shaping the shoreline and features such as sandy beaches and spits predominate. In major estuaries where the tidal range exceeds 4 metres tidal landscapes such as salt marshes and mud flats predominate. Between these two extremes the coastal landscape has elements of both regimes. Erosion in estuaries reaches a maximum at mid-tide when flows are strongest both inwards and outwards, reaching a maximum where these flows are constricted into narrow channels. Deposition, by contrast, reaches a maximum at high and low tide when the water is relatively still. The result is a tendency to sort materials, depositing the finest particles at the inner and outer extremities of the tidal reach to form salt marshes and mud flats, and concentrating the coarsest in the intermediate areas as gravel and sand banks.

Tidal flows also govern the pattern of channels in estuaries and inlets. They alternate daily from quiescence at high and low tide to strong flow at mid-tide both coming in and going out. This twice-daily attack and retreat contrasts with the unidirectional outflow of rivers. It shows up when comparing their plan views. Near the coast tidal channels are wider and have a greater meander wave length than do fluvial (river) channels of comparable discharge. On the other hand, both width and length diminish much more rapidly towards the innermost reach of the tide than do fluvial channels over the same distance between mouth and source (see **Figure 16.7**). These differences seem to be related to the tendency of tidal channels rapidly to disperse the energy of the water during ebb and to concentrate it during inflow times. Evidence for former tidal channels (called *tidal palaeomorphs*) in inland situations can be deduced from these characteristic features on maps and aerial photographs and from the presence of marine fossils or salt deposits. They have been noted in the channels leading inland from the Wadden Sea in northern Netherlands, and in the network of sandy levees in the English Fens, called *roddons*, surviving from Roman times. These are sinuous ridges of light-coloured silty material, originally the tracks of creeks. They are bounded by raised banks, extending from the main river estuaries far inland into the peat country.

Figure 16.11: Mountainous shoreline of submergence (after Strahler page 533). Note the drowned appearance as evidenced by the absence of cliff formation and beaches and the truncation of the drainage streams before they can form depositional areas.

Changing Sea Levels

Shorelines are called neutral if sea level is static. But usually there is evidence that they have risen or fallen in the past, giving *shorelines of submergence* (**Figure 16.11**) and *shorelines of emergence* (**Figure 16.12**) respectively. Where both have occurred, the shorelines are called compound.

Shorelines of submergence result from a drowning of the coastal topography. Valleys become inlets, hills become peninsulas and islands. The drowned valleys are called *rias* and the whole a *ria shoreline*. Where the coastal regions have been glaciated, a *fjord coastline* results. *Fjords* have steeper walls, deeper water, and usually penetrate further inland than rias. They also have submerged *rock bars* across the mouth where the glacier originally debouched into the sea. They are common in the northern hemisphere as a result of the rise in sea levels following the melting of the Pleistocene ice sheets.

Shorelines of emergence, by contrast, present a coastline against the former sea floor. Their elevation steepens the gradients of streams draining from the land, rejuvenating them and causing them to incise gullies. If the previous shoreline was static

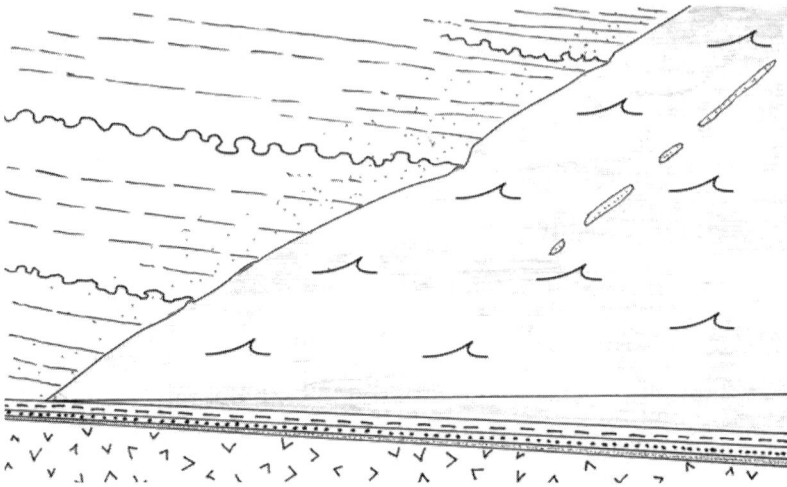

Figure 16.12: Shoreline of emergence: coastal playa type (after Strahler page 533). The land area is an alluvial plain due to former deposition in the sea. Its emergence above sea level has left it an even plain but with slow-flowing meandering streams developing to drain it to present sea level.

Figure 16.13: Beach cusps (after Lobeck, 1939, page 356). These are triangular accumulations of sand and gravel regularly placed along the beach with the apex of each pointing down towards the water. They only occur in bays and are thought to be deposited by down-beach wash following the alternation of waves entering the bay past each of the two encompassing headlands.

for a long time the uplifted and exposed surface will contain layered deposits of clays, silts, sands, and gravels. These tend to give a low, smooth coastal plain, gently sloping seawards and bounded by a simple, even, shoreline with many areas of shallow water offshore. Old beaches and wave-cut benches become elevated into *raised beaches*, where beach materials are still clearly evident and *coastal terraces* where they are not. Old cliffs are isolated inland and tend to recline (decrease in gradient) with the passage of time, as can be seen, for instance, along the Pembrokeshire coast between Pendine and Laugharne (**Figure 2.14**).

A *compound shoreline* shows evidences both of past submergence, such as coastal terraces and cliffs, and of past emergence such as raised beaches and rejuvenated gullying.

Micro-forms

Superimposed on larger beach landforms are smaller features induced by wave action. They include *beach cusps, ripples, transverse ribs, rhomboid ripples,* and *anti-dunes.*

Beach cusps (**Figs 16.3 & 16.13**) are low, regularly spaced crescent-shaped rises in the beach materials with their 'horns' pointing seawards. The coarsest material is in the horns. They always occur in groups, more commonly in shingle than in sand. They may vary in size from a few centimetres to several metres. They seem to originate from minor rises between adjacent cones of swash due either to an alternation of onshore waves from different directions, or to *rip currents*. The latter are narrow, often strong, horizontal circulatory currents with seaward flows typically spaced at regular intervals along the coast. Some may involve combinations of these.

Cusps, micro-ridges on the beach parallel to the shoreline, and arcuate patterns resembling the effects of a shovel swinging in a semicircle, are almost invariably composed of shingle rather than sand.

Ripples (**Figure 16.3**) are transverse sand ridges not more than a few centimetres high with the steeper sides facing down-current. They are associated with relatively slow moving water and increase in size with increasing current speed. Transverse ribs are bars formed of stones or other coarse debris in the sand. They originate because each obstacle generates a hydraulic jump on its upstream side and a cascade on its downstream side. The steeper the bedslope (i.e. the overall slope of the ground) the shorter their wavelength. Rhomboid ripples are interlocking diamond-shaped rises with their steeper sides facing up-current. They occur particularly on intertidal sandbars and seem to be due to the intersection of two directions of flow separated by an acute angle. Anti-dunes are of ripple size but differ from ordinary ripples in that the steeper side faces up-current. They are created by relatively rapid water flow and are often due to backwash on beaches. They tend to have sinuous parallel crests.

Coral reefs

In the warm shallow water off tropical and subtropical coasts colonies of polyps build reefs which expand outwards and build upwards from the sea floor. They often line coasts for many miles and surround islands. They consist of white spongy limestone. Waves swirl and break over them. The polyps thrive best and generate reef expansion at its outer edges where oxygen, calcium carbonate and food are most abundant.

CHAPTER 17

Wind-formed deposits

Wind Action

Wind-formed deposits are found mainly in three types of area: arid, periglacial and coastal, because these are the places where loose detritus accumulates in the relative absence of vegetation.

Wind moves loose materials when it is above a threshold velocity of about 4 metres per second (9 miles per hour). The amount and type of movement depends on the size of particles. Sands and silts move because they are lighter than stones but less cohesive than clays. In general, sand particles between about 2 and 0.05 millimetres in diameter move by creep and saltation (jumping) along the ground but are too heavy to be raised by normal eddy currents. They rise nearly vertically but fall downwind at a low angle. The faster the wind the lower the impact angle.

Silt particles (0.002-0.05 mm in diameter), on the other hand, once entrained, are kept continuously aloft and can be carried long distances from their source areas in deserts and around the margins of ice sheets. They tend to be dropped as loess in semi-desert regions where they are fixed by rain and trapped by vegetation. In such areas they mantle the landscape, sometimes to considerable depth e.g. in northern China and the Mississippi Valley.

On a world scale the accumulations of windborne sand fall into three quite distinct size range classes without distinct intermediates: ripples, *dunes*, and *draa* (**Figure 17.1**). It seems that ephemeral local winds form the first, seasonal and cyclonic winds the second, and uninterrupted Trade and zonal wind

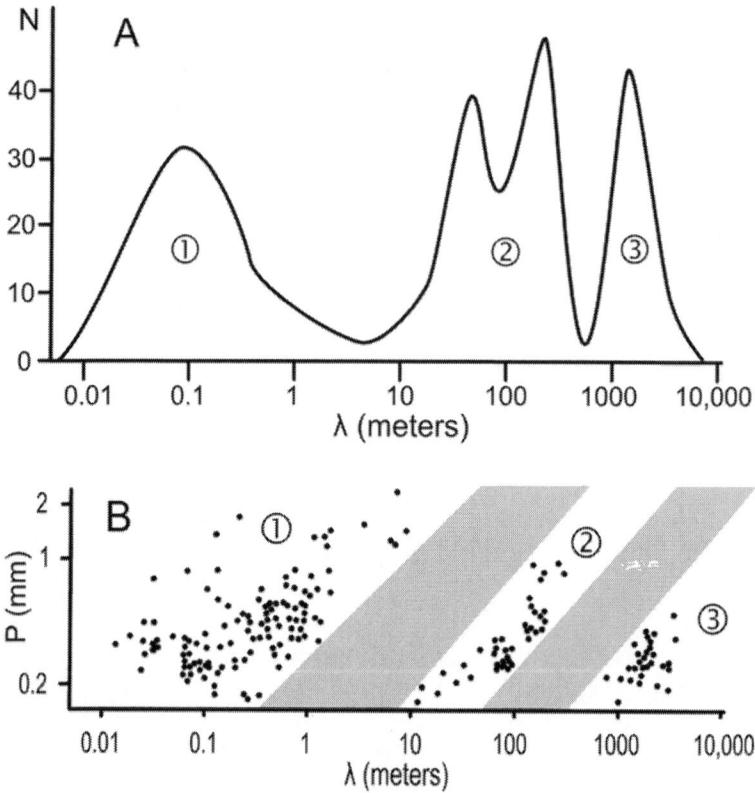

Figure 17.1: In wind dominated arid areas surface material is formed into waves with repeated peaks and valleys. The distance between successive peaks is called the wavelength (λ) and varies from a few cm for surface ripples to several km for draa. They occur together with the slopes of draa being made up of smaller undulations which in turn are made up of ripples. Figure 17a is a graph showing the relationship between the wavelength of bedforms (λ) and an index of the relative proportion of ground area they cover (N). Small bedforms have an upper size limit of about 1 metre (1), dunes have a limit of about 600m (2), and draa of about 6km in extent (3). Figure 17b is a scatter plot of λ versus an index of grain size (P - the coarse 20 percentile) for aeolian bedforms. No transitional forms occur in the grey bars between the three groups (1), (2) and (3) which correspond to ripples, dunes and draa. (after Wilson 1972 page 193).

patterns the third. The resulting features are geographically repetitive and are called bedforms. Their size generally increases with the size of particles which they contain. The reason that the wind forms distinct sand hills separated by bare places, rather

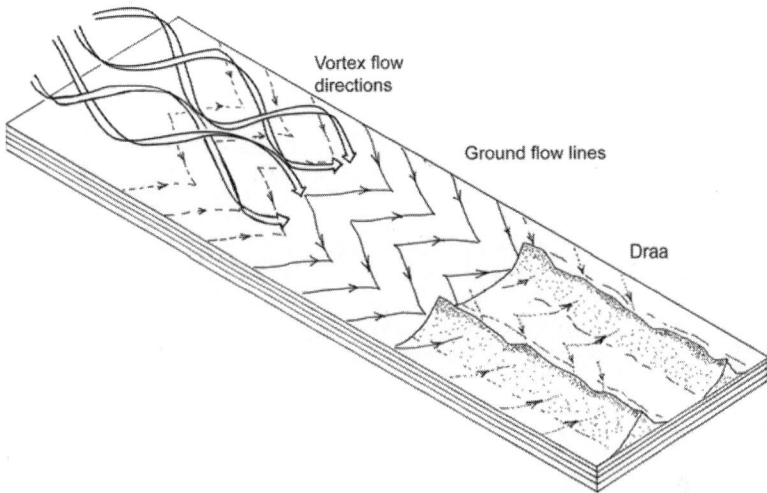

Figure 17.2: Draa dunes formed by 'helical flow'. The wind moves forward in parallel streams with alternating directions of twist. The result is to sweep parallel avenues and deposit the swept material into long dunes between them (after Cooke and Warren, page 291).

than a continuous sand blanket, is that the air does not move forward as a sheet but as a row of adjacent parallel streams, each having a corkscrew-like helical motion with every alternate one having an opposite direction of twist. The effect is to sweep parallel avenues and deposit the swept material in long rhythmically peaked dunes between them (**Figure 17.2**).

Small bedforms

Rotary turbulence in the lowest wind causes surface *ripples* (**Figure 17.3**). These are straight or sinuous micro-ridges transverse to the wind direction 25 cm-3 m apart and 2.5-15 cm high. They are closely analogous to those formed by the sea on beaches. Their size and spacing increase with the coarseness of the particles. They are normally asymmetrical with the steeper slopes facing downwind. Once started they are perpetuated by the fact that saltating grains fall mainly on the windward side of the ripples, so building them up. Ripple length increases generally in the range 2.4-12 cm. Above this size they tend to disappear.

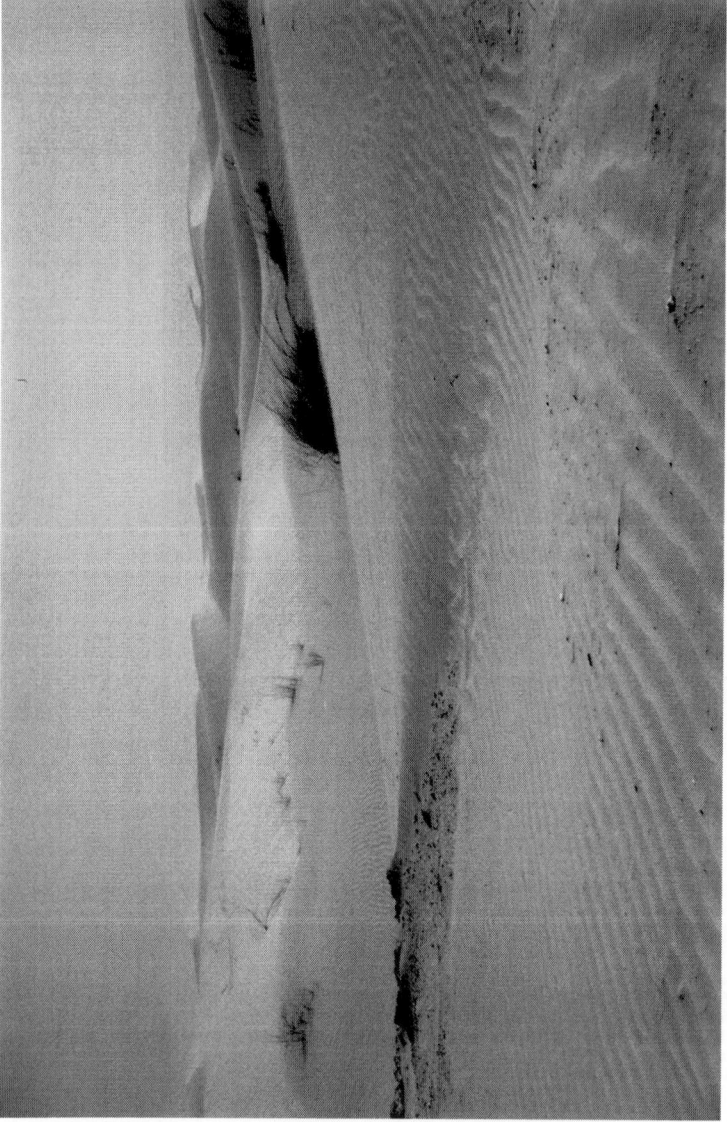

Figure 17.3: Foreground: wind surface ripples; background: draa in Libyan desert.

Figure 17.4: Wind hummock.

Figure 17.5: Nebka or shrub-coppice dune: height c1 metre.

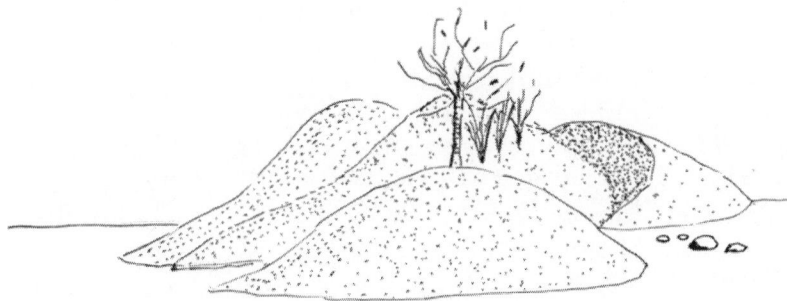

Figure 17.6: Rebdou: height c3 metres.

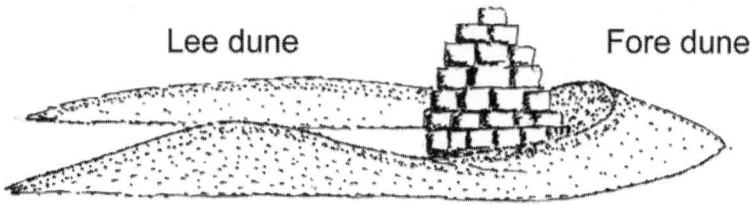

Figure 17.7: Obstacle dunes with the wind from the right: height 1-2 metres. The width and height of the obstacle determine the amount of deposition around it.

Small dune-like features

Wind-formed features larger than ripples but less than about 3 metres in height are small *lee-dunes*. They are are usually caused by obstacles, mainly plants. The five distinguishable types in increasing order of size are: *hummocks* (**Figure 17.4**), *nebkas* (**Figure 17.5**), *rebdous* (**Figure 17.6**), *obstacle-dunes* (**Figure 17.7**), and *sand trains*, sometimes called *windraces*. Hummocks are symmetrical sand accumulations trapped around single plants and are common where there are vegetation tussocks. Nebkas are symmetrical, up to a metre high, and trapped in the lee of small shrubs, boulders, or other obstacles of similar size. They tend to be half - egg - shaped with the stoss end facing up-wind and a tail at the downwind end. Rebdous are similar to nebkas, except that they are asymmetrical and larger, being generally 2-3 metres high. The *shrub-coppice dunes* or *elephant-head dunes* of the American West are of this type.

Obstacle dunes occur where wind speed is slowed around obstacles such as small hills, buildings, or tree clumps. Their size and form depend on the sand supply, the strength of the wind, and the size of the obstacle. They are either climbing dunes on the windward side or falling dunes on the lee side. Under constant strong winds off a beach or other source of sand, they can accumulate to considerable height on the windward side of hills. The obstacle-dune has two parts. That upwind of the obstacle is the *foredune*, that downwind, the *lee-dune*. If the obstacle is steep-sided and seriously interrupts the windflow, the foredune tends to be somewhat crescentic and separated from the obstacle by a small dead-air space. If the obstacle is more

Figure 17.8: 'Smoking dunes', Libya. The apex of the dune receives the strongest wind. This makes it the point of maximum sand receipt and export.

streamlined, the foredune will tend to pile up on its windward side. Dunes completely enveloping the obstacle which caused them are called *zemoul* (singular *zemla*). They indicate an abundant sand supply and moderate winds. When the wind blows parallel to an elongated obstacle such as a wall or hedge, the helical flow pattern will tend to leave a sand-free corridor between the dune and the obstacle.

Sand trains, called in French *trainées de sable*, are long low tails of sand only a few centimetres high deposited in the lee of small obstacles by short-lived high winds sweeping across flat areas.

Dunes

Medium-sized dunes usually range in height between about 3 and 30 metres. Their size and shape depend on sand supply, wind regime and underlying topography. In general, they are best developed downwind of abundant sources of loose sand and on relatively low-lying tracts in the landscape. They generally increase in size away from their source areas up to a certain maximum above which they do not grow. They can be transverse, longitudinal, compound, sheet-shaped, or built round an obstacle. It is sometimes possible to see the driven sand coming over the crest of the dune, - a phenomenon called 'smoking dune' (**Figure 17.8**).

Small dunes are generally 3-5 metres in height. As in larger dunes their surfaces are almost invariably an assemblage of *pack faces* on windward exposures where sand is being added and *slip faces* on lee sides where it is falling over a crest to repose at its angle of rest. Pack faces tend to be firmly packed and rippled, slip faces soft and unstable.

Transverse dunes

Transverse dunes result from a constant moderately strong unidirectional wind acting on a limited sand supply. The characteristic form is the crescentic *barkhan* (**Figure 17.9**) which can have a number of variants, such as the "half-moon" or "broken egg" forms (**Figure 17.10**). When large it shows 3 distinct curves on its slip face. Such barkhans are given the French term *trèfle*. When barkhans grow too big to move or develop as a whole they may become double with three forward horns. When the sand supply is somewhat greater they coalesce laterally into *W-shaped dunes* (**Figure 17.11**). The barkhanoid embayments are called *zirat* (singular *zire*) and the forward promontories, *linguoid dunes* (**Figure 17.12**).

Figure 17.9: Barkhan, Libya

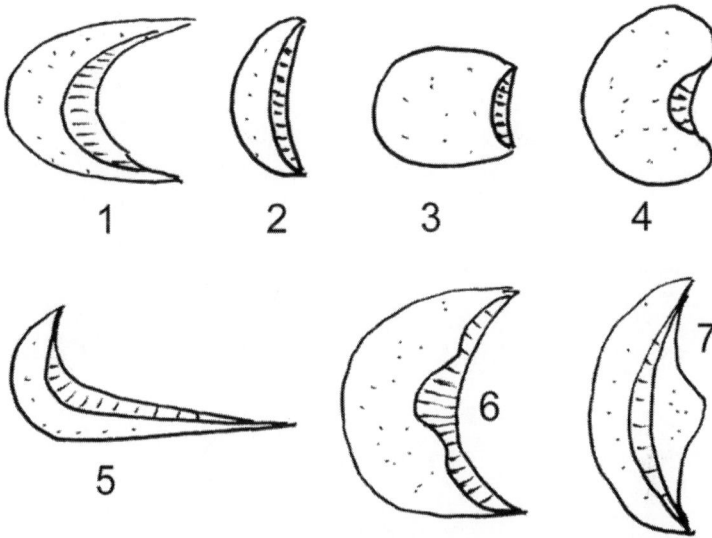

Figure 17.10: Types of simple transverse dunes: 1 normal barchan; 2 half-moon; 3 broken egg; 4 rounded; 5 elongated; 6 trèfle; 7 dune with small sand heap due to wind reversal.

Figure 17.11: Barkhans: single and in chains. Note W-shaped dunes (after Cook and Warren page 294).

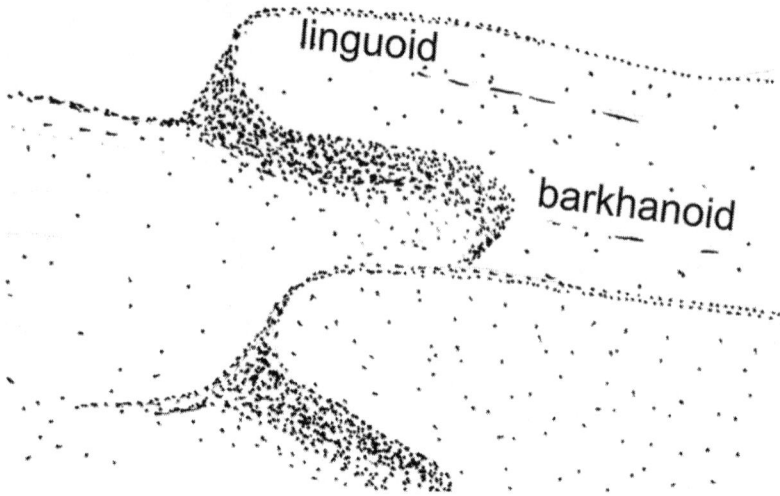

Figure 17.12: Linguoid and barkhanoid dunes.

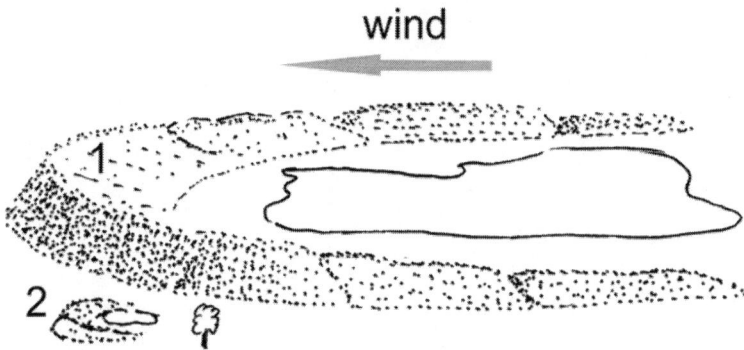

Figure 17.13: Large parabolic dune (1) and small lunette (2) (after Cooke and Warren page 288).

Particles of silt, clay and salts blown out of dry lakes sometimes accumulate immediately downwind of them, where they have the appearance of reversed barkhans with the horns pointing upwind. This is probably because the amount of material available is greatest opposite the centre of the lake. They often become fixed by vegetation, and are known as parabolic dunes or *lunettes* (**Figure 17.13**).

Longitudinal dunes

The common freestanding longitudinal dune is known as the *seif* (**Figure 17.14**) after the Arabic word for sword, although names such as *whaleback, sandridge,* and *linear dune* are sometimes used. The seif is formed by relatively strong winds operating on a limited sand supply and oscillating between two similar directions. Seifs, like draa, usually have S-shaped curving crests. When this sinuosity is marked they are called *sigmoidal dunes*. These sometimes run parallel, separating and joining in an anastomosing fashion. Two seifs sometimes join into one to form a *tuning fork dune* with the prongs of the Y pointing up-wind. These are common in the Simpson Desert of Australia. Also like draa, seif dunes are usually in parallel approximately equidistant groups, due to the same corkscrew-type wind action. In southwestern USA parallel linear dunes are called *wind-row* dunes.

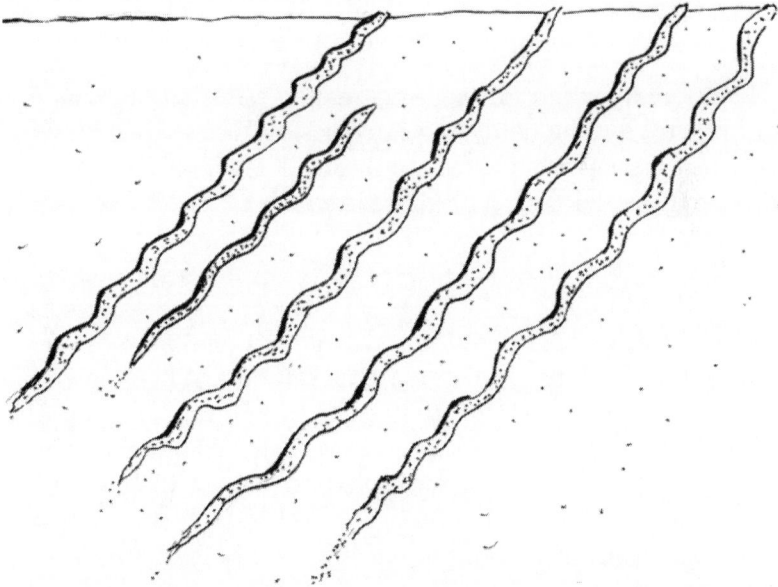

Figure 17.14: Seif dunes on a plain. Wind is from top right. Seif dunes are formed by relatively strong winds operating on a limited sand supply and oscillating between two similar directions. They have s-shaped curving crests and usually run in parallel approximately equidistant groups due to the corkscrew-like wind action (see Fig 17.2).

Figure 17.15: Wind-formed landscape with: 1 ghrouds; 2 demkhas; 3 draa; 4 barkhans; 5 W-shaped dunes; 6 slouk (after Cooke and Warren page 289).

Occasionally small linear dunes, called *slouk* (singular *silk*) (**Figure 17.15**) overlie fields of other sorts of dunes while being oriented at a somewhat different angle. They indicate a past change in the dominant wind direction.

Longitudinal dunes may also be formed partly by selective deflation. The most characteristic types are *hooked dunes*, *parabolic dunes* (**Figure 17.13**), *dunes en rateau* (rake-shaped) and *windrift dunes*. All have in common a concave slope facing the wind which has blown material out of it. When viewed from above such dunes appear cup shaped and often multiple. Hooked dunes are similar but with one limb longer than the other. Dunes en rateau have multiple hooks so as to resemble rakes from the air.

Compound dunes

Where the sand supply is abundant and the winds squally and uncertain in direction, there may be a disorganized dune assemblage called *aklé* (**Figure 17.16**). This consists mainly of wavy transverse ridges of alternating forward linguoid dunes

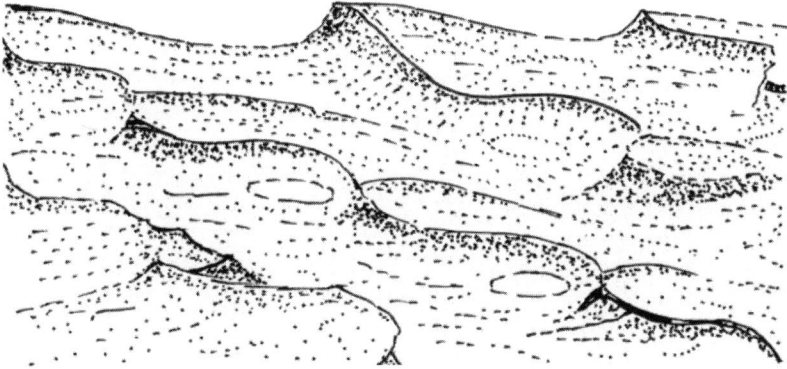

Figure 17.16: Aklé dunes: a disorganized assemblage where sand supply is abundant and winds squally and uncertain in direction. They consist of wavy transverse ridges of alternating linguoid dunes joined by recessed barkhanoid dunes (after Cooke and Warren page 288).

joined by recessed barkhanoid dunes. Consecutive ridges are sometimes arranged in echelon so that the overall appearance can be that of a rectangular reticule. Sometimes the zirat are individually aligned but the totality is not, giving the appearance of a rough sea. Inter-dune hollows, called *fulji* (singular *fulje*), are normally circular or elliptical with their steepest walls on the windward side. Where the alignments of the individual dunes are more regular and have more sand-free passages, the area is called *ansguié* (plural *ansguiat*). In areas of slightly higher rainfall, dunes are rounded and stabilized by vegetation, the higher points being called *nebbiés*. Where a *nebbié* is surrounded by smaller domal dunes, the group is called a *medden*.

Sand Sheets

In many areas windblown sand forms a level sheet. This is especially characteristic of surfaces over disintegrating sandstone, at the outer edges of areas of larger dunes, or around playas. Such sand sheets are seldom extensive, often rippled, and support a little scattered vegetation.

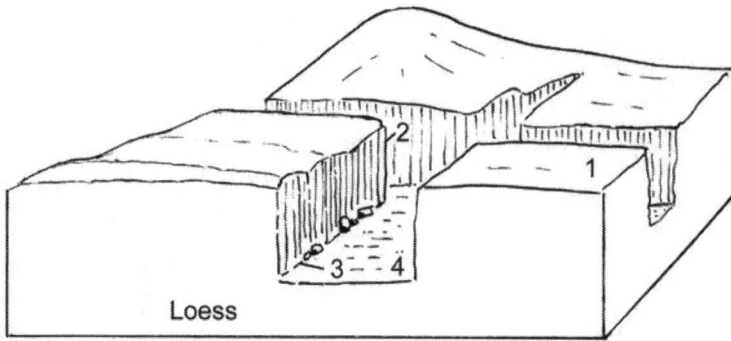

Figure 17.17: Typical loess-covered area. Loess deposits cohere because the small size of particles gives a large overall surface area and their platy shape causes them to interleave. A typical loess landscape shows: 1 undulating top; 2 vertical cliffs; 3 heaps at cliff foot; 4 flattish valley bottoms. The cliffs and valley bottoms may be up to tens of metres in size.

Loess

Loess deposits are soft, unstratified, homogeneous, calcareous, porous, usually yellowish in colour, and often contain the shells of land snails (**Figure 17.17**). They can mantle the landscape to a depth decreasing exponentially away from the source of their materials. The small size of particles gives them a large overall surface area resulting in a strong tendency to cohere. This is increased by their platy shape, which causes them to interleave. The result is that cut slopes in loess tend to be stable at relatively steep gradients, and an eroded loess landscape will typically have an undulating surface cut by deep vertical-sided gullies with relatively level bottoms. Material which falls from the gully sides tends to form piles at the foot without causing the slopes to recline.

Wind-deflated Surfaces

In arid areas, sand and silt are deflated wherever the wind arrives uncharged with sediment. Their removal lowers the level of alluvial land in valleys and basins. It explains the hollowing of playas and dayas. The wind also removes fine particles from the interstices of surface stones and gravel, leaving these as lag deposits. This process increasingly armours the surface so that a maturely deflated surface has a complete cover of stones (see background of **Figure 17.3**). When these are relatively large as over rock outcrops, the result is a stony *hamada* (see **Figure 4.2**). When they are gravel-sized and the surface resembles that of a garden path, it is called *reg* in the western Sahara, *serir* elsewhere.

Wind-abraded Surfaces

Wind-borne sand erodes by blasting exposed stones, forming them into *ventifacts*. Fixed stones facing a roughly constant wind direction are facetted on one side at right angles to the wind and are called *einkanter*. When these are rolled over so that more than one side suffers wind-born abrasion they become *dreikanter*.

An interesting feature of deserts is the way glass can be frosted by wind scour and take on a violet colour on exposure to bright sunlight for a number of years. This was first reported in bottles long abandoned in the sun outside bars in the Australian outback. It has become a tourist attraction and such bottles command good prices.

At a larger scale, the sand blast scours upstanding landforms into yardangs (**Figure 17.18**). These normally have rounded upwind faces and long downwind-pointed tails. If they are composed of bands of unequal strength, a resistant caprock may be considerably undercut because of differential abrasion by wind of lower less resistant beds, such forms are called *zeugen*. An extreme case is the *mushroom rock* formed when the underlying material is eroded into narrow pedestal supporting a wider fragment of more competent rock. Wind abrasion also locally excavates caves and windows and even natural arches.

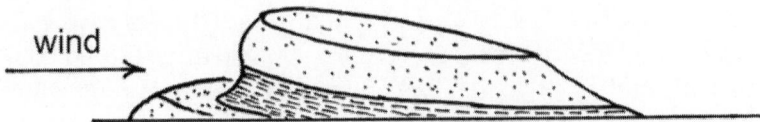

Figure 17.18: Yardang: side view. Yardangs are aligned in the direction of prevailing winds. The windscour causes the upwind face to be rounded and the downwind extension to take the form of a ridge. The erosion of the lower bed makes this into a zeugen.

Figure 17.19: Parallel draa in the area between Saudi Arabia and Yemen. The crests are about 2 km apart (NASA satellite photograph).

Draa

Draa are the largest type of sand bedform. They range in height from several tens of metres to over 400 metres. They are the component sandhills of *ergs*, sometimes called *sand seas*. They occupy 1/4 to 1/5 of the land surface of arid areas. In Arabia, for instance, they cover 300,000 out of 1,600,000 square kilometres of which 230,000 are in the Empty Quarter (Rub al-Khali). Since they are most extensively developed in the Sahara and western Asia, many descriptive terms come from these areas.

Draa tend either to be isolated hills or long parallel ridges (**Figure 17.19**). They sometimes have intermittent peaks ar-

Figure 17.20: Ghourd, Libya. Ghourds are found in extremely arid areas, notably the Sahara, and can reach 450 metres in height. Their peaks are the centre of radiating 'edges' and they are surrounded by lower dunes. They tend to be approximately equidistant, each one focussing the wind energy and sand supply in its immediate area.

ranged in a grid pattern with bare ground appearing between them. Where the peaks focus a number of crests and are star-shaped or pyramidal they are called *ghroud* (singular *ghourd*) (**Figure 17.20**), where rounded, demkhas. Isolated ghroud are aligned with seasonally alternating winds, longitudinal draa with unidirectional trade winds. Examples of both can be seen in the Grand Erg Occidental and Grand Erg Oriental in Algeria. Summits are formed of smaller aklé dunes.

When draa become fixed by vegetation in areas where rainfall has increased since their formation, they lose the sharp crests and steep slopes of actively moving sand and become undulating and rounded. This is exemplified by the *qozes* of western Sudan.

In deserts with constant high winds, sand may be carried over relatively high obstacles. The highest ghroud in the world, topping 300 metres, appear to be in the Issaouen-en-Ighargharen basin, Algeria, whose enclosing hills are hundreds of metres higher than its floor. This basin is far too small to have itself provided such vast quantities of sand.

At this large scale wind-erosion phenomena are dominated by grooves and large deflation basins. Grooves cut in rock plains may be half to one kilometre wide and 500 metres to 2 kilometres apart. They are, for instance, widespread on the southeastern flanks of the Tibesti Mountains in the Sahara, where they cover over 90,000 square kilometres, and where their aeolian origin is proved by their regular parallelism and exact accordance with the direction of the dominant Trade Winds.

Deserts also contain large closed basins relatively clear of sand. The best known are the Qattara Depression and the oases in Egypt. Although their origin is debated, it seems most likely that their hollowing is due to long continued deflation of loose debris washed into their centres by erosion from their sides.

CHAPTER 18

The hand of humans

Introduction

Very few landscapes are untouched by human hands and many are profoundly modified. Fortresses, roads, canals, and agricultural terraces have existed for millennia. The last century has seen accelerated change: vast quarries and mining tips, motorways, coastal protection works, artificial lakes, and reclamations of land from the sea.

In Britain today vast quantities of building aggregate are excavated annually, more than 10 times the amount taken at the start of the 20th century. There are today something like 10,000 abandoned mine workings in Britain. There is considerable ground subsidence in mining areas and rapid erosion around coasts. The burning of fossil fuels has made rain more acid and accelerated land degradation in unmeasured ways.

Artificial hills

Probably the earliest, and still among the biggest, constructions altering the landscape are religious mounds. These include the ziggurats of Mesopotamia, the pyramids of Egypt, the vast pyramidal hill presumed to contain the tomb of the Emperor Qui Shi Huang near Xian in China, and similar structures in Mexico and Chile. The largest in Europe is Silbury Hill (Wiltshire) (**Figure 18.1**) which is 40 metres high. Stone Age *long barrows* and Bronze Age *round barrows* (**Figure 18.2**) are conspicuous features in the landscape less for their size than because they were located so as to break the horizon when viewed from a distance. The Vikings seem to have been responsible for mounds with

Figure 18.1: Silbury Hill is 40 metres high and the largest artificial mound in Europe, visible from Avebury Circle about 1 km away. Its purpose is still unknown. (after Collins in Hippisley Cox page 11).

Figure 18.2: Beacon Hill from Seven Barrows, near Lambourn, Berkshire. Beacon Hill has artificial military ramparts 1 km in circumference. The nearby Seven Barrows are excavated burial tumuli in a conspicuous position (after Collins in Hippisley Cox page 103).

terraced edges slightly raised above the surrounding country, called *thing mounts*. Examples can be seen at Fell Foot Farm in Little Langdale in the Lake District and in the Isle of Man.

Military Works

Military excavations and buildings have always been conspicuous but have changed profoundly through history. In Britain the earliest are probably the Neolithic *causewayed camps*, rings of earthworks through which run causeways. Larger examples are the Iron-Age *hilltop forts* such as Uffington Castle in

Figure 18.3: Maiden Castle near Dorchester was an ancient fort, enlarged with a ditch and rampart in the early Iron Age, and then transformed by natives in Roman times to enclose almost 40 hectares. This involved building an inner rampart towering more than 15m above the bottom of its fronting ditch, using limestone brought from over 3 km` away. The entrance defences were both complex and formidable and it became an inhabited township. (after Collins in Hippisley Cox page 75).

Berkshire, and Badbury Rings, and Maiden Castle in Dorset (**Figure 18.3**). Some forts are ramparts, protected on one side like Bredon Hill (Worcestershire). The Romans introduced large forts and watchtowers, most clearly seen on Hadrian's Wall. The Saxons left few traces although some earthworks such as Offa's Dyke in the Welsh Marches and the Devil's Dyke in Cambridgeshire date from their period. On a smaller scale many castles, churches and even houses were surrounded by moats which interposed a stretch of water between them and potential attackers.

The Norman invasion led to the rapid building of defensive earthworks of the motte-and-bailey type, notably along the Welsh border. This is a quickly constructed earth and timber fortification consisting of a circular artificial mound surrounded by a ditch with a separately defended bailey or courtyard beside it. The medieval period saw the construction of fortified towns and large stone castles all over Europe. Avignon, Carcassonne, Chester, and York are examples of the former, Pierrefonds, Amboise, Harlech, Corfe Castle, and Bodiam of the latter. Some large Crusader castles such as at Kerak in Jordan and Krak des Chevaliers in Syria, date from the same period. In Britain, moated settlements are even more widespread than castles. Over 5,500 have been identified, many of which were constructed between 1200 and 1335 AD. The moats were probably used as a source of water and a protection against wild animals as well as for military defence.

The arrival of gunpowder decreased the height but increased the intricacy and sophistication of defensive walls and moats. The commonest design came to be a *bastioned trace* whereby a city was surrounded by a wall with a series of arrowhead-shaped projections. Berwick-on-Tweed, Metz, and Antwerp are examples.

Agriculture

Agriculture has frequently been the cause of erosion, usually due to the removal of covering vegetation, especially forest. Trees trap much rainfall and "breathe out" much moisture, reducing the water that reaches the ground. Tree roots bind the soil and the annual leaf fall feeds the soil fauna of worms and other small creatures. These play a key role in maintaining a good soil structure, which ensures good permeability and through drainage.

Tree clearance ends these benefits. Soil creeps downhill and gullies develop. The loosened materials rapidly accumulate in valley bottoms and lake basins. In Europe this process is widespread and has been going on for more than 5,000 years. In Britain the farm pattern was already largely established by Iron Age and Roman times. In some low lying areas it was on a recognizable grid pattern. Recent removal of hedgerows and the ploughing up of ancient pastures has further reduced vegetation cover.

Wind erosion often follows vegetation clearance. Sandy and peaty soils are especially vulnerable. They dry out in early summer when there can be spells of dry windy weather, and are then exposed to the burrowing and nibbling of rabbits. The overwhelming of the village of Santon Downham in the Breckland of East Anglia in 1668 seems to have been from this cause. Soil-blows on the drained organic peats of the Fenlands have become commoner in recent years. This seems to be due to the reduced protection resulting from the removal of hedge-

Figure 18.4: Lynchetts near Bishopstone, Wiltshire (after Collins in Hippisley Cox page 143).

rows, the use of artificial fertilizers rather than farmyard manure, and the increased cultivation of sugar beet which requires a loose soil and leaves the land relatively bare in early summer.

Ploughing began about 6,000 years ago when Neolithic farmers cleared the forest with flint hand axes and used a simple plough to cultivate wheat and barley. The earliest conspicuous evidence in Britain is from *strip lynchets* (**Figure 18.4**). These are small terraces formed by ploughing along the contour of hillslopes in such a way that the plough cuts a sharp line on the uphill side of a field and the disturbed soil moves downhill leaving a well-marked little scarp at the upper edge and a fall at the outer edge of the material dislodged. The lynchets thus appear as strips of land, often relatively flat, that stand out distinctly from the slopes above and below. They can differ in height from the adjacent hillslope or a neighbouring lynchet by anything from a few centimetres to 8 metres. They may occur as staircases of parallel strips, especially on steep slopes in southern England. Medieval lynchets tend to be longer than their prehistoric forbears and can be up to 200 metres long.

Celtic fields tended to be small and squarish partly because farmers ploughed them in both directions. Although less than 1 acre in size they are sometimes surrounded by great banks. They covered thousands of square miles on chalk downland, moorland and other terrain that escaped medieval cultivation. They still survive in Celtic areas of Wales, Scotland, and Ireland.

Another consequence of ploughing is the ridge-and-furrow pattern common in some parts of the English Midlands where the fields were approximately rectangular. This seems to have resulted from the practise of minimising plough distances by working outwards from a single furrow. Every time the farmer ploughed he went up one side of the ridge and back the other in such a way that his mouldboard plough turned both furrows inwards towards the ridge crest. The ploughing unit forming one ridge was called the *selion* which was usually a furlong in length (220 yards = 200m) and 11 yards (10m) in width. Selions generally ran up and down slope so as to make drainage easier. After completing one strip of a few furrows the farmer would start another until he had covered the whole field. Across the end of a series of selions there was often a *headland ridge* or *balk*. The whole system was arranged in a block of approximately rectangular shape.

Field boundaries are often very ancient. All over southern Dartmoor are long parallel stony banks called *reaves*. They are typically about 100 yards (91m) apart, run across country for miles, and are divided by cross reaves at irregular intervals. They were fully in use during the Bronze Age and tell a story of country planning on a gigantic scale. Similar patterns of sub-division of equal antiquity occur as hedges in the Saints area of northeast Suffolk. Some in the low lands of the eastern counties show a grid pattern. There is evidence from Ireland that such boundaries may date from the Neolithic.

Locally the Romans also divided the country into a grid pattern called *centuriation*. One square had 710 metres on a side (equal to 20 Roman *acti*). There are many examples of this around the Mediterranean, for instance in France and Italy. In parts of Tunisia one pattern of centuriation overlies an earlier one with a slightly different orientation. An example of centuriation in Britain occurs around Holme-next-the-Sea in Norfolk, aligned along the Roman road of Peddars Way. Such patterns also occur in the Ripe and Chelvington areas of West Sussex and the Dengie peninsula of Essex (**Figure 18.5**), although these may have originated before the Roman conquest (Rackham page 119).

Figure 18.5: Iron Age Roman field grid on the low clay land of the Dengie peninsula near to the Essex coast. This has remained agricultural because of its fertility. It is clearly visible from the air (after Rackham , plate XIII).

Bridges also go back many centuries. For instance there was an ancient bridge from the Dengie peninsula across the Crouch estuary from North to South Fambridge (called Fanbruge = 'fen bridge' in Domesday Book). A principal Roman road would have crossed here, but there has been no bridge since the Middle Ages. The tides of the Crouch now sweep 1/4 mile wide between these bridgeheads, although the Ordnance Survey remembers a ferry. There was a mediaeval bridge 3 miles upstream at Hullbridge which fell down in the 17th century and has not been replaced. The lowest 20th century bridge, Battlesbridge, is 5 miles above the Anglo-Saxon bridge.

Peat diggings

Humans have brought different changes to basin peats from those they have brought to mountain peats. In the former, change has largely come from deflation after drying and from its extraction for fuel.

Draining peaty, organic soils lowers the water table. This leaves the peat to be dried, oxidized, and sometimes blown away. In parts of East Anglia this has lowered ground levels by as much as 4 metres since 1848. The lowering of the peat areas in the Fens has exposed roddons. Because of their relative elevation these have become the site of buildings and roads.

The lower courses of rivers Yare and Waveney in east Norfolk flow sluggishly in flat marshy valleys where basin peat has developed. They include the open water called the Broads. These were thought for a long time to be a natural feature but are now known to be medieval peat diggings which have since become flooded. The cutting of basin peat is now mainly confined to Ireland where it us used both for domestic fuel and in power stations. Most of that country's deeper raised bogs have already disappeared.

Human-induced change to mountain peat, also called blanket peat, has mainly come from erosion. Peat absorbs much water and controls stream flows but its low cohesion makes it inherently unstable. This instability increases as it becomes saturated with rainwater, eventually leading to *bog slides* and *bog bursts*. These leave bare, unstable, and unvegetated surfaces which soon become eroded. Gullies

Figure 18.6: Cattle track which has become a 'holloway' to Borough Hill Camp, near Daventry (after Collins in Hippisley Cox page 186).

cut back into the peat mass itself. The eroded peat is washed downslope and accumulates against objects in its path, such as walls, fences, and even shrubs.

The rate of erosion of mountain peats has increased in the past two or three centuries due to human activities such as sheep grazing, burning, ditch digging, and the recreational tramping of tourists. Foresters sometimes cut parallel ditches to drain the ground so as to plant trees. This digging process breaks the protective sod and often exposes the underlying materials to accelerated erosion.

Engineering works

Engineering works include road and railway cuttings and embankments, canals, harbours, pipelines, and drains. All require excavation and removal of the detritus. Drains have had some unforseen consequences. For instance, by accelerating river flows they may alter the locus of deposition. This can cause ponding or flooding in new places.

Roads have always been conspicuous in the landscape. The Romans developed an integrated network in Britain and some of their roads still survive. Their major roads typically had 2 side ditches between which was a raised embankment (Latin *agger*). The ditches were 24m (80 feet) wide, the agger 12 metres (40 feet) wide and 1 metre (3 feet) high, covered with a 6m (20 foot) width of gravel. More recently roads have influenced terrain through the occurrence of *holloways*. These are unpaved routeways roads deeply sunk in ravines due to wear and erosion (**Figure 18.6**).

Cities grow on top of their own waste. It has been estimated that on average they dump about 30 centimetres of debris per century. The City of London, for instance, rests on an average of 3-5 metres of debris while some ancient cities such as Erbil and Mosul in northern Iraq are built on mounds over 10 metres high.

The use of rivers

We control rivers for water supply, navigation, flood protection, and irrigation. Rivers are the basis of life in many arid areas. The Nile, Indus, and Tigris-Euphrates, for instance, have been

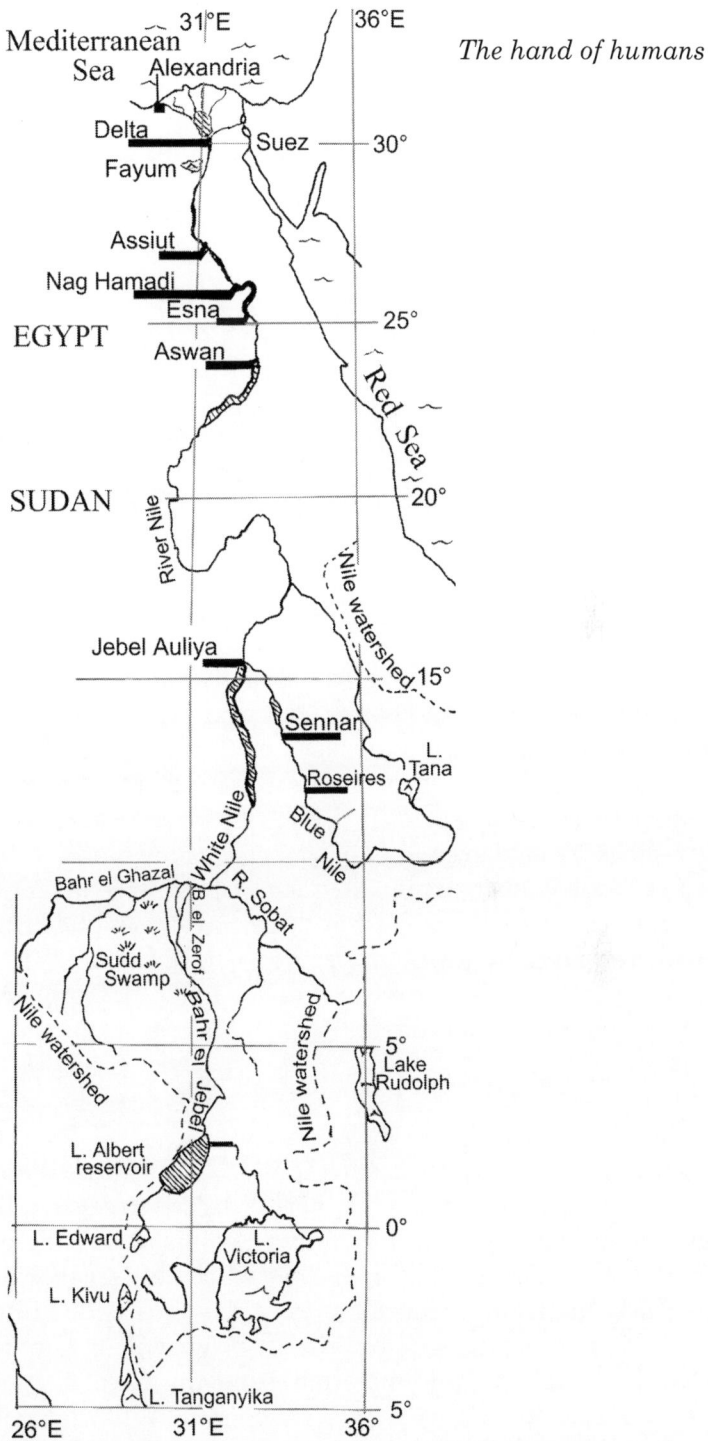

Figure 18.7: Irrigation changes to terrain in the Nile system. Heavy horizontal lines show dams and barrages; oblique shading shows induced lakes (after Ludwig:main map).

controlled to irrigate wide areas for millennia. In recent times chains of dams have multiplied their coverage (**Figure 18.7**).

Drainage ditches in Holland and the English fens make agriculture possible in areas that otherwise would be water-logged. These works have many less obvious effects. One example of such cutting is Vermuyden's Old Bedford River cut from Earith towards the Wash in the early 17th century to prevent the flooding of the Ouse. It had to be followed by the New Bedford River parallel to it to allow the land between them to be inundated when the Ouse was in spate. This had the incidental effect of emptying the lower reaches of the former Ouse and allowing the Cam, a former tributary, to provide the main flow. Dams have the opposite effect on their downstream channels, causing them to become smaller than their natural predecessors because they are not subjected to the same peak flows.

Urban developments also affect drainage channels. Impermeable surfaces of cement, tile, and tarmac increase the rapidity of runoff and the presence of storm drains and sewers causes peak flows to be more intense and more rapidly attained than elsewhere. When such flows are diverted into natural channels, their extra volume erodes the banks. Mining waste can also affect stream channels. The nineteenth century zinc mines of mid-Wales and the china clay mines of Cornwall have given large volumes of waste which have silted up rivers and their estuaries. Erosion from ploughed-up agricultural land may have a similar effect.

Mining

Mining has been practised since prehistoric times. Neolithic men dug pits in the chalk of southern England to extract flints. The best known site is Grimes Graves (Norfolk), but there are also flint mines at Cissbury on the South Downs near Worthing. They also quarried greenstone from the Prescelly Hills in Pembrokeshire, augite-granopyrite from near Penmaenmawr in North Wales, volcanic materials from the Great Langdale District of the Lake District, and porcellanite from Cushendall (Antrim).

During the Bronze Age copper was mined in Scotland and Wales, tin in Cornwall, and lead in the Mendips, Shropshire, and

Figure 18.8A

Figure 18.8B

Figure 18.8: The effects of opencast coal mining near Shef.eld, April 2003. A) scraping and slumping on a hill slope, B) tabulation and the effects of roads.

the Pennines. It became a major industry in northern England in the seventeenth and eighteenth centuries. Spoil heaps, lateral excavations into hillsides called *adits* , and channels, called leats, for bringing water along contours for ore washing and drainage have remained as visible traces.

Iron working in Britain is likewise ancient. Miners extracted ores by grubbing the surface and digging pits. They then smelted them in shallow pits. The remains are called bloomeries and usually contain heaps of partially roasted ironstone, cinder and slag. The Romans worked sites in the Weald and the Forest of Dean. The Wealden mines flourished and were still being exploited in the eighteenth century when they were associated with the famous hammer ponds. These were created by damming narrow valleys to provide water to work the iron

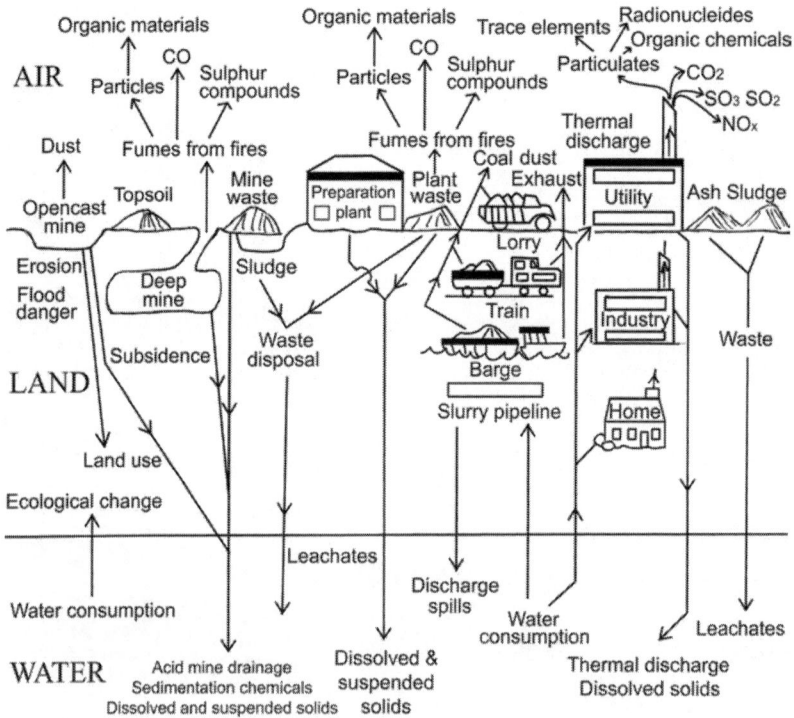

Figure 18.9: The whole process of coal extraction and its effect on land, water and air. Note particularly the amount of gases and particulates emitted into the air, the massive solid waste added to the land surface, and the volume and nature of the liquid drainage.

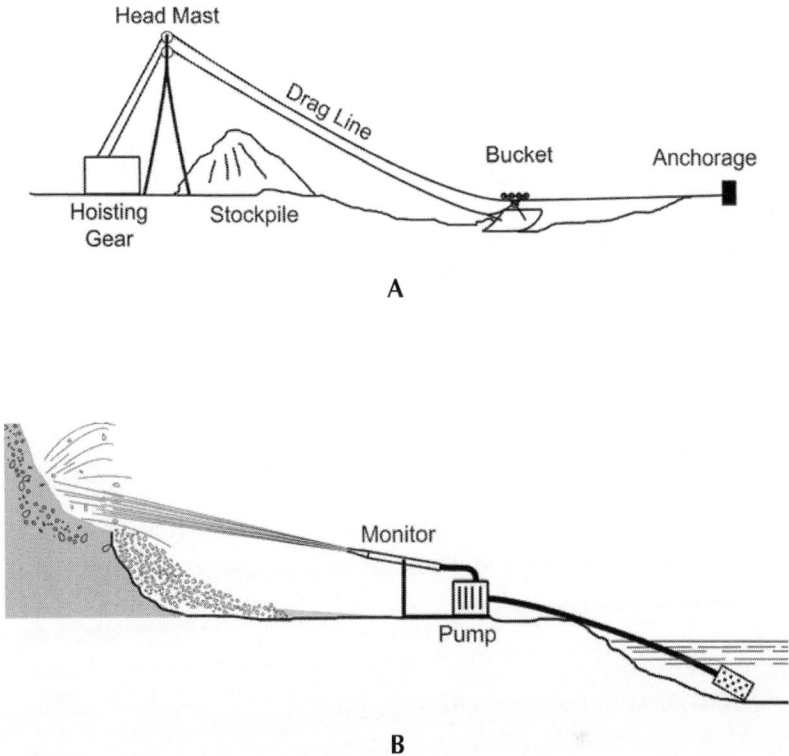

Figure 18.10: Land alteration caused by gravel extraction: A) extraction-and stock pile; B) water blast and gravel slump (after Sand and Gravel Association of Great Britain (SAGA) pages 73 and 78).

mills. They are sometimes be arranged in a series and appear like a string of beads from the air.

The Industrial Revolution changed the whole scale of mining operations. Great pits and spoil tips resulted from the mining of china clay in Cornwall and brick clays from Peterborough and in Bedfordshire. Huge quarries provided ballast for the railways, limestone for iron furnaces, and building stones for the great cities. Opencast coal mining has left deep quarries across the countryside (**Figure 18.8**) and has lowered the ground in many areas. It has sometimes left the land surface in a series of stepped levels called *tabulation*. Deep mining of coal has left vast coal tips (though many are now carefully landscaped) and areas of subsidence, especially in South Wales, Durham, and the Potteries. **Figure 18.9** shows the whole process of coal working and its effects on land, water, and air.

Excavation in one place leads to accumulation in another. Mining yields pit heaps. It is estimated that there are about 2 billion tonnes of shale waste in pit heaps around Britain. Such accumulations can cause disasters when they collapse. In Aberfan (South Wales) a disastrous failure occurred when part of a 180-metre high coal waste tip slid and flowed downhill. The tip had not only been constructed with a very steep un-stable slope but had also been located where water naturally seeped out of the ground. Saturation of the lower parts of the waste lowered its strength and triggered failure.

Mining causes land subsidence. It can produce depressions, disrupt surface drainage, and cause permanent lakes. It damages buildings. The amount of subsidence can be considerable, 12 metres being reported in parts of Staffordshire. Subsidence has also followed salt extraction. Salt is mined by dissolving it and pumping out the brine. In parts of Cheshire this has led to subsidence and the formation of lakes. The extraction of surface gravel leaves large pits or slumping slopes (**Figure 18.10**).

Human actions on coasts

Coasts are protected by beaches of shingle and sand. When industry extracts these materials for building purposes, natural processes will partly replace them. The reason for this is that a great deal of loose material around the British coasts is of some antiquity. During the cold glacial phases so much water was stored in the ice caps that world sea levels fell. This exposed much of the continental shelf as dry land. Rivers continued to bring down a load of detritus and piled it on this shelf. When the ice sheets melted the sea level rose and it combed up much of this detritus to form beaches. So long as sea level continued to rise it washed more shingle up on to the shore. But by the end of the Ice Age the sea had more or less reached its present level and this combing process ceased. Many places, therefore have finite resources of shingle.

The most serious consequence of shingle removal is prob-ably the loss of protection for the foot of cliffs. A classic example of the effect of human activity is Folkestone Warren which lies to the east of the town. Chalk overlies clay along the coast. Originally the cliff was protected by a beach at its foot which was fed by shingle and sand washed along the coast from the west, but the construction of large harbour works at Folkestone in the nineteenth century blocked this movement and the Warren area became depleted of beach material. This exposed the cliffs to undercutting by the sea and there was a series of failures, the largest in 1915 disrupting the main railway line from Folkestone to Dover. The area has now had to be stabilized with massive sea-defence works.

The removal of beach material has also threatened low ground near to the sea. The village of Hallsands in South Devon is an example. This was a fishing settlement on a small platform between sea and cliff. The beach material was extracted to a considerable depth and used to extend the naval dockyard at Devonport. The extraction was so extensive that the village was seriously exposed to wave damage, and destroyed by a severe easterly storm in 1917.

Groynes, piers, and breakwaters alter coasts. They interrupt longshore sediment movement, causing accumulation on the up-current side but permitting erosion down-current where the coast is starved of its sand supply. Seaford (Sussex), for instance, suffers from the construction of the Newhaven breakwater and Lowestoft suffers from the building of the pier at Gorleston. The building of a breakwater at the western extremity of Chesil Beach has stopped the westerly movement of fine shingle by longshore drift, accumulating it on the eastern side of the breakwater. As a consequence, erosion, demanding expensive counter-measures, has started on the western side of the breakwater.

Artificial sea walls and dykes protect the coast from erosion, and are often needed in coastal cites to prevent flooding, as in London, Amsterdam and New Orleans. They can, however, leave areas of coarse rubble when destroyed by heavy seas. This occurs especially when their foundations are undermined by backwash. In some places, such as the Lleyn Peninsula (North Wales) a stable protection is provided by dumping boulders along the top of the beach large enough to blunt wave attack and resist movement except in the most severe storms. The outstanding example of human control of the sea is the Netherlands where over the centuries a large area of country has been reclaimed and enclosed behind dykes (**Figure 18.11**).

Coastal areas which can be easily enclosed into shallow lakes have been used as salt evaporating pans. These can be seen today around the Mediterranean and at the southern end of the Dead Sea. There were salt pans in the Pevensey Levels of Sussex in the Middle Ages.

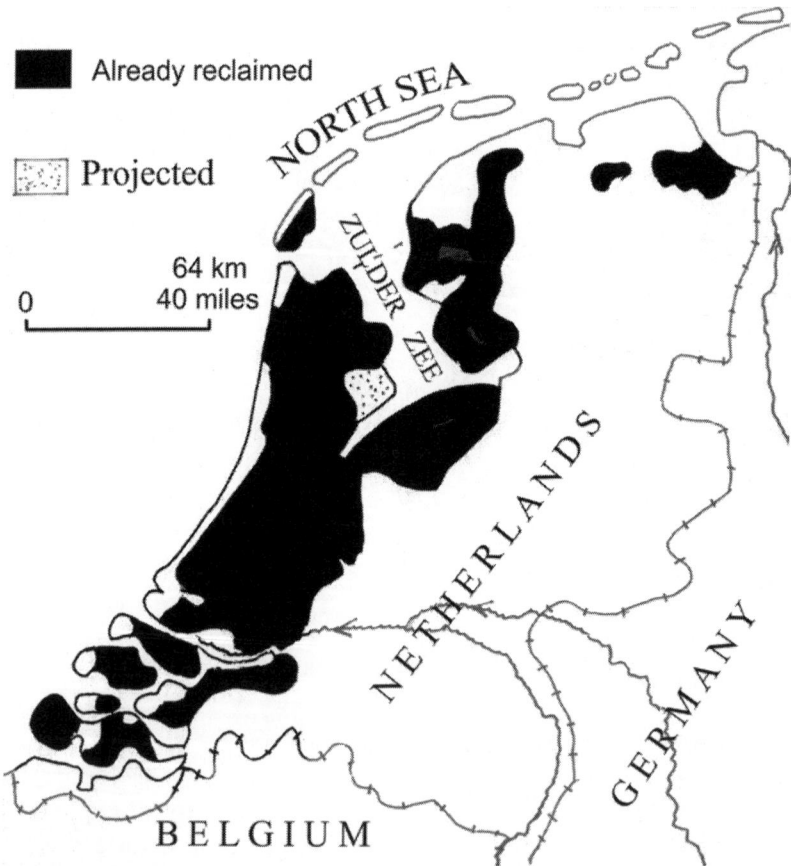

Figure 18.11: Land reclaimed from the sea in the Netherlands (after Shackleton page 172).

Changes to coastal dunes

The pressures of tourist motoring and tramping decrease the stability of coastal dunes because of their reduction of the vegetation cover. This is undesirable since dunes provide protection from wave attack and their topographic variety often makes them valuable habitats for unusual plants and wildlife and provides sites for animals to burrow. In Britain, various methods have been used to protect them. Mechanical means include wood and wire fences and nylon netting but these can be unsightly. The best solutions have included the careful planning of car parks and access routes, the protection of vegetation against grazing, and the encouragement of sand trapping, using binding plants. One such is the sea buckthorn *Hippophae rhamnoides* but this can become rampant, necessitating control. The most spectacular long-term success has been with the Culbin Sands in northeastern Scotland. The estate and its house had to be abandoned in 1694-5 after submergence under blowing sand, the largest such area in the country. But since 1921 the Forestry Commission have stabilized the area by planting it with conifers.

Aerial pollution

The accelerated burning of fossil fuels (coal, oil, and gas) over the last few decades has made rainfall markedly more acidic and corrosive (see **Figure 18.9**). pH values of less than 3 have been recorded by comparison with just under 6 for unpolluted rain. This accelerates the weathering of many rocks. Also the increasing amount of carbon dioxide in the air has increased the amount of carbonic acid in surface waters. This makes a dilute acid which is highly effective in dissolving limestone.

Global Warming

It is perhaps apposite to end this book about the earth with some comments on global warming. The prevailing popular belief is that human production of greenhouse gasses has been the primary cause of global warming since 1750 and must be curtailed, at enormous expense. It is based on the main man made greenhouse gas, carbon dioxide.

It is beyond dispute that carbon dioxide levels are rising. The atmospheric carbon dioxide proportion stood at 381 parts per million (ppm) in 2005 compared with 280 ppm in 1850. Some of this increase is due to man's burning of fossil fuels. Carbon dioxide emissions have increased from around 1 billion metric tons per year in 1950 to 7 billion in 2005. It is also clear that the earth is getting warmer. The annual average temperature of the earth's surface has increased from 13.77°C in 1880 to 14.43°C in 2005, the hottest year on record at the time of writing. Data are available on carbon dioxide levels and global temperatures dating back for hundreds of thousands of years and the two have gone up and down together.

This association between temperature and carbon dioxide is an undisputed fact and suggests a causal link between them but the popular belief about the origin and cure of global warming overstates the evidence in one critical respect: what is cause and what is effect? Close inspection of carbon dioxide levels and temperatures shows that carbon dioxide tends to follow rather than lead temperature changes suggesting that it is the effect and not the cause of temperature rise. There is a ready explanation for this as rising temperature reduces the solubility of carbon dioxide in sea water, releasing it into the atmosphere. Another problem with the popular theory is that while human activity may have driven up carbon dioxide levels in recent times there is nothing obvious to have driven them up and down in the past other than changing temperature. On the other hand there is an obvious driver for changing temperature: changing solar activity.

While the cause of rising global temperatures remains uncertain, global warming is beyond dispute and has a number of effects on the landscape. Most obvious are the retreat of ice cover in peri-polar regions and the advance of desert in the arid tropics. Melting ice that is already floating does not influence sea levels but melting of ice on land does. The huge ice cover on the land masses of Antarctica and Greenland if melted could raise sea levels by potentially more than 60 metres. This adds to the effect of thermal expansion of the existing oceans. Most predictions of the rise in sea levels over the coming century are between 10cm and 1m. Other effects are less certain but an increase in extreme weather seems likely with accelerated erosion in localised areas. These changes have occurred repeatedly in the past and bring benefits such as improving agricultural productivity in cooler regions as well as the costs of expanding deserts in the arid tropics and flooding of low-lands.

References

Bloom, Arthur L (1978) Geomorphology: a Systematic Analysis of Late Cenozoic Landforms, Prentice-Hall Inc., Englewood Cliffs, New Jersey 07632.

Cleare, John (1979) The World Guide to Mountains and Mountaineering, Webb & Bower, Exeter.

Collard, Roy (1989) The Physical Geography of Landscape, Unwin Hyman, London.

Cooke, R U & Warren, A (1973) Geomorphology in Deserts, B T Batsford, London.

Cox see Hippisley

Davis, William Morris (1906) the sculpture of mountains by glaciers. Scot Geog. Mag. vol.22 pp76-89, Geog. Essays pp 617-634.

Evans, I O (1953) The Observer's Book of British Geology, Frederick Warne & Co., London.

Geikie, Sir Archibald (1887) The Scenery of Scotland, MacMillan & Co., London.

Geikie, Sir Archibald (1903) Textbook of Geology, Macmillan & Co., New York.

Gorshkov, G and Yakushova, A, translated from the Russian by A. Gurevich (1967) Physical Geology, Mir Publishers, Moscow

Hills, E. Sherbon (1965) Outlines of Structural Geology, Methuen & Co Ltd, London.

Henoch, W E S (1960) String-bogs in the Arctic 400 miles north of the Arctic Circle, Geographical Journal, London, vol 126, 335-339

Hippisley Cox, R (1944), illustrated by W W Collins, RI, The Green Roads of England, Methuen & Co Ltd, London.

Holmes, Arthur (1956) Principles of Physical Geology, Thomas Nelson & Sons Ltd, London.

King, Philip B and Schumm, Stanley A (1980) The Physical Geography (Geomorphology) of William Morris Davis, compiled, illustrated, edited, and annotated, Geo Books, Geo Abstracts, Norwich, England. Labhart, T P & Decrouez, D (1995) Geologie de la Suisse, Ott verlag, Thun.

Leopold L B and Wolman M G (1957) River channel patterns - braided, meandering, and straight, US Geological Survey Professional Paper 282-B, 39-85.

Lobeck, Armin Kohl (1924) Block Diagrams and other graphic methods used in Geology and Geography, John Wiley & Sons, New York (c/o Dept of Geol & Geog, University of Wisconsin, Madison, Wis.)

Lobeck, Armin Kohl (1939) Geomorphology, McGraw-Hill, New York.

Longwell, Chester R, Knopf, Adolph, and Flint, Richard F (1941) Outlines of Physical Geography, John Wiley & Sons, Inc., New York.

Ludwig, Emil (1936) The Nile: the Life Story of a River, from the Source to Egypt, translated by Mary H Lindsay, George Allen & Unwin Ltd., London.

Mabbutt, J A (1977) Desert Landforms, MIT Press, Cambridge, Massachusetts

Mackinder, H J (1902) Britain and the British Seas. William Heinemann, London.

Mainguet, Monique (1972) Le Modelé des Grès, 2 vols, Institut Geographique National, Paris

Miller, Terence (1953) Geology and Scenery in Britain, B T Batsford, Ltd. 4 Fitzhardinge Street, Portman Square, London W1.

Muir, Richard (1943) The new reading the landscape: fieldwork in landscape history, University of Exeter press, Exeter.

Oard, Michael (2004) Pediments formed by the Flood: evidence for the Flood/post-Flood boundary in the late Cenozoic, Technical Journal (TJ) Answers in Genesis, Acacia Ridge D.C., Queensland, Australia.

Rackham, O (1986) The History of the Countryside; the Full Fascinating Story of Britain's Landscape, J M Dent & Sons, London & Melbourne.

Saint Joseph, J K S (1966) The Uses of Air Photography: Nature and Man in a new Perspective, John Baker Publishers, Ltd, London

Sand and Gravel Association of Great Britain (SAGA) (1967) Pit and Quarry Textbook, Macdonald & Co, Ltd, London

Schmid, Max (1985) Iceland - The Exotic North, Iceland Review, Reykjavik, Iceland

Shackleton, Margaret Reid (1947) Europe: A Regional Geography, Longmans, Green & Co., London.

Shepherd, Walter (1952) The Living Landscape of Britain, Faber and Faber, 24 Russell Square, London WC1.

Small, R J (1970) The Study of Landforms, Cambridge University Press, Bentley House, 200 Euston Road, London NW1

Soil Survey Staff, U S Department of Agriculture (1960) Soil Classification: A Comprehensive System, 7th Approximation, Washington, DC.

Sparks, B W (1962) Geomorphology, Longmans, 48 Grosvenor Street, London W1.

Stephens, N (ed) (1990) Natural Landscapes of Britain, Cambridge University Press, Cambridge.

Stonehouse, Bernard (1982) The Aerofilms Book of Britain from the Air, Weidenfeld & Nicolson, London

Strahler, Arthur N (1969), Physical Geography, John Wiley & Sons, New York and London

Sweeting, Marjorie M (ed) (1981) Karst Geomorphology, Hutchinson Ross Publishing Company, Stroudsburg, Pennsylvania.

Thornbury, W. D (1954) Principles of Geomorphology, John Wiley & Sons, Inc., New York.

Wainwright, Alfred (1966) The Western Fells, Westmorland Gazette, Kendal.

Ward, Colin R (ed) (1984) Coal Geology and Coal Technology, Blackwell Scientific Publications, Melbourne, Australia.

Whittow, John B (1984) The Penguin Dictionary of Physical Geography, Penguin Books, London.

Wilson, I G (1972) Aeolian bedforms - their development and origins, Sedimentology, 19:173-210.

Index

A

D

S

Printed in Great Britain
by Amazon.co.uk, Ltd.,
Marston Gate.